Lecture Notes in Artificial Intelligence 7472

Subseries of Lecture Notes in Computer Science

LNAI Series Editors

Randy Goebel
University of Alberta, Edmonton, Canada
Yuzuru Tanaka
Hokkaido University, Sapporo, Japan
Wolfgang Wahlster
DFKI and Saarland University, Saarbrücken, Germany

LNAI Founding Series Editor

Joerg Siekmann
DFKI and Saarland University, Saarbrücken, Germany

T0226330

Martin Atzmueller   Alvin Chin
Denis Helic   Andreas Hotho (Eds.)

Modeling and Mining Ubiquitous Social Media

International Workshops
MSM 2011, Boston, MA, USA, October 9, 2011
and MUSE 2011, Athens, Greece, September 5, 2011
Revised Selected Papers

 Springer

Series Editors

Randy Goebel, University of Alberta, Edmonton, Canada
Jörg Siekmann, University of Saarland, Saarbrücken, Germany
Wolfgang Wahlster, DFKI and University of Saarland, Saarbrücken, Germany

Volume Editors

Martin Atzmueller
University of Kassel, Knowledge and Data Engineering Group
Wilhelmshöher Allee 73, 34121 Kassel, Germany
E-mail: atzmueller@cs.uni-kassel.de

Alvin Chin
Nokia Research Center, Mobile Social Networking Group
Beijing 100176, China
E-mail: alvin.chin@nokia.com

Denis Helic
Graz University of Technology, Faculty of Computer Science
Inffeldgasse 21a, 8010 Graz, Austria
E-mail: dhelic@tugraz.at

Andreas Hotho
University of Würzburg, Data Mining and Information Retrieval Group
Am Hubland, 97074 Würzburg, Germany
E-mail: hotho@informatik.uni-wuerzburg.de

ISSN 0302-9743   e-ISSN 1611-3349
ISBN 978-3-642-33683-6   e-ISBN 978-3-642-33684-3
DOI 10.1007/978-3-642-33684-3
Springer Heidelberg Dordrecht London New York

Library of Congress Control Number: 2012947587

CR Subject Classification (1998): I.2.6, H.3.3-5, H.3.7, H.5.3-4, H.2.8, H.2.4, C.5.3,
I.2.9, K.4.4, K.6.5, C.2.4

LNCS Sublibrary: SL 7 – Artificial Intelligence

Typesetting: Camera-ready by author, data conversion by Scientific Publishing Services, Chennai, India

Printed on acid-free paper

Springer is part of Springer Science+Business Media (www.springer.com)

Preface

The emergence of ubiquitous computing has started to create new environments consisting of small, heterogeneous, and distributed devices that foster the social interaction of users in several dimensions. Similarly, the upcoming social web and social media technologies also integrate the user interactions in social networking environments.

Social media and ubiquitous data are thus expanding their breadth and depth. On the one hand, there are more and more social media applications; on the other hand, ubiquitous sensors are becoming part of personal and societal life at larger scales, as well as the accompanying social computational devices and applications. This can be observed in many domains and contexts, including events and activities in business and personal life.

Understanding and modeling ubiquitous (and) social systems require novel approaches and new techniques for their analysis. This book sets out to explore this emerging space by presenting a number of current approaches and important work addressing selected aspects of this problem. The individual contributions of this book focus on problems related to the modeling and mining of ubiquitous social media. Methods for mining, modeling, and engineering can help to advance our understanding of the dynamics and structures inherent to systems integrating and applying ubiquitous social media. Specifically, we advance on the analysis of dynamics and behavior on social media and ubiquitous data, e.g., concerning human contact networks or anomalous behavior. Furthermore, we consider helpful preprocessing methods for ubiquitous data and also focus on user modeling, privacy, and security aspects in order to address the user perspective of the respective systems.

The papers presented in this book are revised and significantly extended versions of papers submitted to two related workshops: The Second International Workshop on Mining Ubiquitous and Social Environments (MUSE 2011), which was held on September 5, 2011, in conjunction with the European Conference on Machine Learning and Principles and Practice of Knowledge Discovery in Databases (ECML-PKDD 2011) in Athens, Greece, and the Second International Workshop on Modeling Social Media (MSM'2011) that was held on October 9, 2011, in conjunction with IEEE SocialCom 2011 in Boston, USA. With respect to these two complementing workshop themes, the papers contained in this volume form a starting point for bridging the gap between the social and ubiquitous worlds: Both social media applications and ubiquitous systems benefit from modeling aspects, either at the system level or for providing a sound data basis for further analysis and mining options. On the other hand, data analysis and data mining can provide novel insights into the user behavior in social media systems and thus similarly enhance and support modeling prospects. In the following, we briefly discuss the themes of these two workshops in more detail.

Social Media Modeling: Social media applications such as blogs, microblogs, wikis, news aggregation sites, and social tagging systems have pervaded the Web and have transformed the way people communicate and interact with each other online. In order to understand and effectively design social media systems, we need to develop models that are capable of reflecting their complex, multifaceted socio-technological nature. Modeling social media applications enables us to understand and predict their evolution, explain their dynamics or to describe their underlying social-computational mechanics. Much of the user interfaces used to access social media are difficult to use and do not translate well when shown on a mobile device such as a mobile phone. Moreover, design of social media on mobile devices is significantly different to creating PC-based or Web-based social media applications. Also, social media applications are typically modeled in an application-specific way and therefore there is no standardized method of creating a user interface. User interface modeling in social media can use a wide range of modeling perspectives such as justificative, explanative, descriptive, formative, predictive models and approaches (statistical modeling, conceptual modeling, temporal modeling, etc).

Ubiquitous Data Mining: Ubiquitous data require novel analysis methods including new methods for data mining and machine learning. Unlike in traditional data-mining scenarios, data do not emerge from a small number of (heterogeneous) data sources, but potentially from hundreds to millions of different sources. As there is only minimal coordination, these sources can overlap or diverge in any possible way. In typical ubiquitous settings, the mining system can be implemented inside the small devices and sometimes on central servers, for real-time applications, similar to common mining approaches. Steps into this new and exciting application area are the analysis of the collected new data, the adaptation of well-known data-mining and machine-learning algorithms and finally the development of new algorithms. The advancement of such algorithms with their application in social and ubiquitous settings is one of the core contributions of this book.

Concerning the range of topics, we broadly consider three main themes: communities and networks in ubiquitous social media, mining approaches, and issues of user modeling, privacy, and security.

For the first main theme, we focus on the dynamics and the behavior concerning communities, anomalies, and human contact networks: We consider advanced community mining and analysis methods with respect to both social media and insights into ubiquitous human contact networks. Furthermore, network evaluation and anomalous behavior are discussed. "Integrating Social Media Data for Community Detection" by Jiliang Tang, Xufei Wang, and Huan Liu presents an approach for combining different data sources available in social media (networks) for enhanced community detection. Martin Atzmueller, Stephan Doerfel, Andreas Hotho, Folke Mitzlaff, and Gerd Stumme present related analysis and approaches in "Face-to-Face Contacts at a Conference: Dynamics of Communities and Roles" in a conferencing scenario. Philipp Singer, Claudia Wagner, and Markus Strohmaier discuss the evolution of content and social media in

"Factors Influencing the Co-evolution of Social and Content Networks in Online Social Media." In "Mining Dense Structures to Uncover Anomalous Behavior in Financial Network Data," Ursula Redmond, Martin Harrigan, and Padraig Cunningham discuss the discovery of anomalies in social networks and social media.

Concerning the mining approaches in ubiquitous social media, we distinguish explorative approaches for characterization and description, and mining methods for improving models, e.g., for interpolating ubiquitous data and for modeling temporal effects. For the first dimension, "Describing Locations Using Tags and Images: Explorative Pattern Mining in Social Media" by Florian Lemmerich and Martin Atzmueller presents an explorative pattern-mining approach for characterizing locations using tagging and image data. For the model-oriented techniques, "Learning and Transferring Geographically Weighted Regression Trees Across Time" by Annalisa Appice, Michelangelo Ceci, Donato Malerba, and Antonietta Lanza proposes a spatial data-mining method for geographically weighted regression. After that the paper "Trend Cluster-Based Kriging Interpolation in Sensor Data Networks" by Pietro Guccione, Annalisa Appice, Anna Ciampi, and Donato Malerba describes an interpolation method for ubiquitous data acquired, for example, in the context of pervasive sensor networks.

For user modeling, privacy, and security, we include a simulation approach and the modeling of user interfaces targeting privacy and security aspects. Else Nygren presents a simulation-based approach in "Simulation of User Participation and Interaction in Online Discussion Groups." Next, Ricardo Tesoriero, Mohamed Bourimi, Fatih Karatas, Pedro Villanueva, Thomas Barth, and Philipp Schwarte discuss privacy and security issues in "Privacy and Security in Multi-Modal UIs with Model-Driven Modeling."

It is the hope of the editors that this book (a) catches the attention of an audience interested in recent problems and advancements in the fields of social media, online social networks, and ubiquitous data and (b) helps to spark a conversation on new problems related to the engineering, modeling, mining, and analysis of ubiquitous social media and systems integrating these areas.

We want to thank the workshop and proceedings reviewers for their careful help in selecting and the authors for improving the provided submissions. We also thank all the authors for their contributions and the presenters for the interesting talks and the lively discussion at both workshops. Without these individuals we would not have been able to produce such a book.

May 2012

Martin Atzmueller
Alvin Chin
Denis Helic
Andreas Hotho

Organization

Program Committee

Martin Atzmueller	University of Kassel, Germany
Jordi Cabot	INRIA-École des Mines de Nantes, France
Ciro Cattuto	Institute for Scientific Interchange (ISI) Foundation, Italy
Michelangelo Ceci	Università degli Studi di Bari, Italy
Alvin Chin	Nokia Research Center
Marco De Gemmis	University of Bari, Italy
Wai-Tat Fu	University of Illinois, USA
Denis Helic	KMI, TU-Graz, Austria
Andreas Hotho	University of Würzburg, Germany
Huan Liu	Arizona State University, USA
Ion Muslea	Language Weaver, Inc.
Claudia Müller-Birn	FU Berlin, Germany
Else Nygren	Uppsala University, Sweden
Giovanni Semeraro	University of Bari, Italy
Marc Smith	Connected Action Consulting Group
Myra Spiliopoulou	University of Magdeburg, Germany
Gerd Stumme	University of Kassel, Germany
Christoph Trattner	IICM, TU-Graz, Austria

Additional Reviewers

Abbasi, Mohammad Ali
Aiello, Luca Maria
Canovas Izquierdo, Javier Luis
Martínez, Salvador
Panisson, André
Pio, Gianvito

Table of Contents

Integrating Social Media Data
for Community Detection

Jiliang Tang, Xufei Wang, and Huan Liu

Computer Science & Engineering, Arizona State University, Tempe, AZ 85281
{Jiliang.Tang,Xufei.Wang,Huan.Liu}@asu.edu

Abstract. Community detection is an unsupervised learning task that
discovers groups such that group members share more similarities or
interact more frequently among themselves than with people outside
groups. In social media, link information can reveal heterogeneous relationships of various strengths, but often can be noisy. Since different
sources of data in social media can provide complementary information,
e.g., bookmarking and tagging data indicates user interests, frequency
of commenting suggests the strength of ties, etc., we propose to integrate social media data of multiple types for improving the performance
of community detection. We present a joint optimization framework to
integrate multiple data sources for community detection. Empirical evaluation on both synthetic data and real-world social media data shows significant performance improvement of the proposed approach. This work
elaborates the need for and challenges of multi-source integration of heterogeneous data types, and provides a principled way of *multi-source*
community detection.

Keywords: Community Detection, Multi-source Integration, Social
Media Data.

1 Introduction

Social media is quickly becoming an integral part of our life. Facebook, one of the
most popular social media websites, has more than 500 million users and more
than 30 billion pieces of content shared each month[1]. YouTube attracts 2 billion
video views per day[2]. Social media users can have various online social activities, e.g., forming connections, updating their status, and sharing their interested
stories and movies. The pervasive use of social media offers research opportuni-
ties of group behavior. One fundamental problem is to identify groups among
individuals if the group information is not explicitly available [1]. A group (or
a community) can be considered as a set of users who interact more frequently
or share more similarities among themselves than those outside the group. This
topic has many applications such as relational learning, behavior modeling and

---

[1] http://www.facebook.com/press/info.php?statistics
[2] http://mashable.com/2010/05/17/youtube-2-billion-views/

M. Atzmueller et al. (Eds.): MSM/MUSE 2011, LNAI 7472, pp. 1–20, 2012.

prediction [19], linked feature selection [17] [18], visualization, and group formation analysis [1].

Different from connections formed by people in the physical world, users of
social media have greater freedom to connect to a greater number of users in
various ways and for disparate reasons. In online social networks, the low cost of
link information can lead to networks with heterogeneous relationship strengths
(e.g., acquaintances and best friends mixed together) [24]. Hence, noise and
casual links are prevalent in social media, posing challenges to the link-based
community detection algorithms [12,13,4]. In addition to link information that
indicates interactions, there are other sources of information that indirectly represent connections of different kinds in social media.

User profiles that describe their locations, interests, education background,
etc. provide useful information differing from links. For example, Scellato et
al. find that clusters of friends are often geographically close [14]. There are
other activities that produce information about interactions: bookmarking data
implies user interests, frequency data of commenting on their friends homepage
suggests the strength of connections. These types of information can also be
useful in finding a community structure in social media.

Figure 1 shows a toy example with two sources, i.e., link and tag information.
Figure 1(a) shows the communities identified by Modularity Maximization [12]

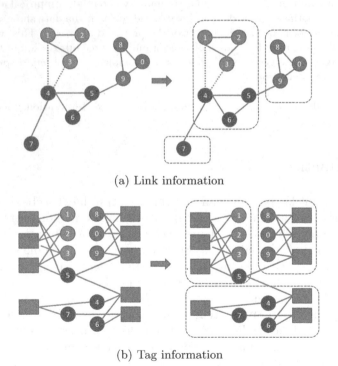

(a) Link information

(b) Tag information

Fig. 1. Community Detection based on Single Source

based on link information. The weak links, i.e., $(1,4)$ and $(3,4)$, make link-based algorithms ineffective. Figure 1(b) shows the results of k-means on the tagging information, which similarly cannot reveal the real community structures. Each source contains noisy but complementary information with other sources. For example, the links $(1,4)$ and $(3,4)$ are weak since they don't share any tagging information.

With these multiple and complementary sources, we ask 1) *could we improve the performance of community detection by combining multiple types of data?* And 2) *how can we integrate data of heterogeneous types effectively?* In this paper, we propose a joint optimization framework to integrate multiple data sources to discover communities in social media. Experimental results on synthetic data and real-world social media data show that the performance of community detection is significantly improved through integrating multiple sources. Our main contributions are summarized below,

- Identifying the need for integrating multiple sources for community detection in social media,
- Proposing a novel framework to integrate multiple sources for community detection and link strength prediction, and
- Presenting interesting findings such as integrating more data sources does not necessarily bring about better performance through experiment design in real-world social media datasets.

The rest of this paper is organized as follows. The related work is summarized in Section 2. The problem of multi-source integration is formally defined in Section 3. An integrating framework is introduced in Section 4, followed by empirical evaluation in Section 5 with detailed discussion. The conclusion and future work is presented in Section 6.

2 Related Work

Community detection algorithms can be divided into three generic categories based on types of data sources used: link-based, link and content-based and interaction-based algorithms. Next we review each category separately.

2.1 Community Detection Based on Links

The study of link-based methods has a long history. It is closely related to graph partitioning in graph theory. For example, one approach to graph partitioning is to find disjoint subgraphs such that cuts are minimized. Since the graph partition problem is NP-hard, it is relaxed to spectral clustering for practical reasons [8]. The concept of modularity to measure the strength of a community structure is proposed in [12]. Since maximizing modularity is NP-hard, a relaxation to spectral clustering is proposed in [23].

A social media user can have multiple interactions and interests, which suggests that community structures often overlap. CFinder [13] is a local algorithm that enumerates all k-cliques and then combines any two cliques if they share $k - 1$ nodes. It is computationally expensive. Evans et al. [5] propose to partition links of a line graph to uncover the overlapping community structure. A line graph can be constructed from the original graph, i.e., each vertex in the line graph corresponds to an original edge and a link in the line graph represents the adjacency between two edges in the original graph. However, this algorithm is memory inefficient, so it cannot be applied to large social networks. EdgeCluster [19] takes an edge centric view of the graph: edges are treated as instances and nodes as features, and can find highly overlapping communities. Some other ways to obtain overlapping communities include soft clustering [11] and probabilistic models [4].

2.2   Combining Link and Content Information

Generative models such as Latent Dirichlet Allocation (LDA) [2] can be used to model links and content via a shared set of community memberships. Erosheva et al. integrates abstracts and references of scientific papers under the LDA framework in document clustering applications [4]. They assume there is a fixed number of categories, each is viewed as a multinomial distribution on words or links. One problem with the generative models is that they are susceptible to irrelevant keywords. [25] proposes a probabilistic model to combine link and content information in community detection with improvement. They first build a conditional model which estimates the probability of connecting node i to node j. Then the membership of a node to a community is modeled on content information and the two models are unified via community memberships. [7] proposes the Topic-Link LDA model that co-clusters documents or blogs and authors. There are two problems with above models: 1) they are designed to model author-emails and author-scientific papers with specific assumptions; and 2) they are not designed to integrate more than two sources as needed for social media.

2.3   Utilizing Interactions beyond Links

Social media users have various types of interactions. Since interactions between users imply their closeness, information of interactions can be important in uncovering groups in social media. A co-clustering framework is proposed in [22] to leverage users' tagging behavior in community detection. It shows that more accurate community structures can be obtained by leveraging the tag information. MetaGraph Factorization (MetaFac) is presented in [6] to extract community structures from various interactions. In [20], the authors propose methods of integrating information of heterogeneous interactions for community detection. Our proposed community detection approach differs from these methods in explicitly integrating tie strength prediction.

3   Problem Statement

Before building the mathematical model, we would like to establish the notations to be used. Following the standard notations, scalars are denoted by low-case letters $(a, b, \ldots; \alpha, \beta, \ldots)$, vectors are written as low-case bold letters $(\mathbf{a}, \mathbf{b}, \ldots)$ and matrices correspond to bold-face upper-case letters $(\mathbf{A}, \mathbf{B}, \ldots)$. $\mathbf{A}(i, j)$ is the entry at the ith row and j^{th} column of the matrix $\mathbf{A}$, $\mathbf{A}(i, :)$ is the ith row of $\mathbf{A}$ and $\mathbf{A}(:, j)$ is the jth column of $\mathbf{A}$ etc. We use $\mathbf{0}_{i \times j}$ to represent a $i \times j$ zero matrix, $\mathbf{I}_k$ to represent a $k \times k$ unit matrix, and $\mathbf{1}_{i \times j}$ represents a $i \times j$ all one matrix. Let $\mathbf{u} = \{u_1, u_2, ..., u_n\}$ be the user set where n is the number of users, and $\mathbf{c} = \{c_1, c_2, ..., c_K\}$ where K communities are to be identified.

Definition 1. Link Matrix $\mathbf{Y}_0 \in \mathbb{R}^{n \times n}$ is the adjacency matrix whose entries represent the connectivity between two users, i.e., $\mathbf{Y}_0(i, j) = 1$ if u_j has a link to u_i, otherwise $\mathbf{Y}_0(i, j) = 0$.

In social networks, the degree distribution typically follows a power law distribution, i.e., most people have a few friends, while few people have extremely many friends. It suggests that the link matrix $\mathbf{Y}_0$ should be sparse. Actually the nonzero entities in $\mathbf{Y}_0$, i.e., the total number of edges or arcs, is normally linear, rather than squared, with respect to the number of nodes in a network. This can be verified following the properties of a power law distribution.

$$p(x) = (1 - \alpha)x^{-\alpha}, x \geq x_{min} > 0 \tag{1}$$

where α is the exponent which often falls between 2 and 3 [10], x is the nodal degree. The expected number of edges is

$$E[\mu^m] = \frac{n}{2} \cdot \frac{\alpha - 1}{\alpha - 2} \cdot x_{min} \tag{2}$$

Definition 2. Affiliation Matrix is denoted by $\mathbf{H} \in \mathbb{R}^{K \times n}$. The jth column of $\mathbf{H}$, $\mathbf{H}(:, j)$, represents the memberships of u_j with respect to K communities so Affiliation Matrix should be *non-negative*.

The diversity of people's interests suggests that people might belong to more than one community. Since the number of communities one belongs to can be upper bounded by his nodal degree, $\mathbf{H}$ should be *sparse*.

Definition 3. Source Matrix is denoted by $\mathbf{Y}_i \in \mathbb{R}^{m_i \times n}(1 \leq i \leq m)$, where m_i is the number of features related to the source i and m is the number of additional sources. If user u_i subscribes to a feature j (e.g., u_i uses the jth tag, or comments on u_j's post), then the corresponding entry is the frequency u_i subscribes to the feature j, otherwise 0.

Source Matrix should also be sparse. For example, one person u_i usually comments on a small part of persons in $\mathbf{u}$. The entities in Source Matrix are not limited to $\{0, 1\}$ since they represent frequencies. The set of m sources is represented by $\mathcal{S} = \{\mathbf{Y}_1, \mathbf{Y}_2, ..., \mathbf{Y}_m\}$.

With the notations and definitions defined, our community detection problem of integrating multiple sources can be stated as follows:

Given Link Matrix $\mathbf{Y}_0$, a set of Source Matrices $\mathcal{S}$, and the number of communities K, compute a sparse Affiliation Matrix $\mathbf{H}$ by leveraging different types of data in social media.

4   A Joint Optimization Framework for Integrating Multiple Data Types

In social media, one person can have multiple activities (e.g., tagging, commenting, etc.). Links contain the static relation between users. It is about one aspect of a user and can be supplemented with additional types of information that reflect interactions of corresponding aspects. For example, tagging data implies personal interests; frequency data of commenting suggests the strength of a connection and so on. Taking into account of different data sources, we investigate how to integrate data of different types in solving the problem of community detection. In this section we begin with a formulation that integrates two data sources before generalizing it to handle multiple data sources.

4.1   Integrating Two Sources

The formation of communities in social media can be explained by the Homophily effect [9]: compared with people outside of the group, users within a group tend to share more commonalities such as forming more connections, interacting more frequently, using similar tags, having similar attitudes, etc. Thus, it is reasonable to assume that people have similar community affiliations in different sources.

Given the link matrix $\mathbf{Y}_0$ and another type of data $\mathbf{Y}_i$, integrating two data sources can be formulated as a joint optimization problem through matrix factorization techniques as follows (2JointMF),

$$\min_{\mathbf{W}_0,\mathbf{W}_i,\mathbf{H}} \|\mathbf{Y}_0 - \mathbf{W}_0\mathbf{H}\|_F^2 + \|\mathbf{Y}_i - \mathbf{W}_i\mathbf{H}\|_F^2$$

$$+ \lambda \sum_{j=1}^{n} \|\mathbf{H}(:,j)\|_1^2 + \eta(\|\mathbf{W}_0\|_F^2 + \|\mathbf{W}_i\|_F^2),$$

$$\text{s.t.}\quad \mathbf{R} = \mathbf{W}_0\mathbf{H} \leq 1_{n \times n}$$

$$\mathbf{R} = \mathbf{W}_0\mathbf{H} \geq 0_{n \times n}$$

$$\mathbf{H} \geq 0_{K \times n} \tag{3}$$

where $\| \cdot \|_F$ denotes the Frobenius norm of a matrix, $\mathbf{W}_0 \in \mathbb{R}^{n \times K}$ and $\mathbf{W}_i \in \mathbb{R}^{m_i \times K}$. The parameter η controls the size of the elements in $\mathbf{W}_0$ and $\mathbf{W}_i$. $\mathbf{H}$ is the Affiliation Matrix, which indicates the memberships of users w.r.t K communities. From the definition of Affiliation Matrix, $\mathbf{H}$ should be non-negative. L_1-norm regularization is widely used for the purpose of achieving sparsity of the solution [21]. In our formulation, L_1-norm regularization is applied to each column of affiliation matrix $\mathbf{H}$ based on the observation that one user is usually involved in a small number of communities. λ balances the trade-off between the sparseness of $\mathbf{H}$ and the accuracy of approximation.

The low cost of link information can lead to networks with heterogeneous relationship strengths [24]. Weak links in online social networks might make link-based community detection algorithms ineffective, as shown in Figure 1(a), and users' multiple interactions indicate the link strengths between users [24] [16].

We use $\mathbf{R}$ to reconstruct the original link matrix and represent strengths of refined relationships between users by considering multiple sources.

Unfortunately, the formulation in Eq. (3) is not concave due to the coupling of $\mathbf{W}_0$, $\mathbf{W}_i$ and $\mathbf{H}$. Thus it is hard to find a global solution for the joint optimization problem. Actually, if we fix 2 components such as $\mathbf{W}_0$ and $\mathbf{W}_i$, the resulting optimization problem for the left 1 component, $\mathbf{H}$, is concave, therefore through computing $\mathbf{W}_0$, $\mathbf{W}_i$ and $\mathbf{H}$ alternatively, we can find a local minimal solution for Eq. (3).

For computing $\mathbf{H}$, we fix components $\mathbf{W}_0$ and $\mathbf{W}_i$ and then develop the following theorem:

Theorem 1. *When components $\mathbf{W}_0$ and $\mathbf{W}_i$ are fixed, the formulation to optimize $\mathbf{H}$ in Eq (3) is equivalent to the following constrained minimization problem:*

$$\min_{\mathbf{H}} \|\mathbf{A} - \mathbf{B}\mathbf{H}\|_F^2$$

$$\text{s.t.} \quad \mathbf{CH} \leq \mathbf{D} \tag{4}$$

where $\mathbf{A}$, $\mathbf{B}$, $\mathbf{C}$, and $\mathbf{D}$ are defined as follows:

$$\mathbf{A} = (\mathbf{Y}_0^\top, \mathbf{Y}_i^\top, \mathbf{0}_{n\times 1})^\top$$
$$\mathbf{B} = (\mathbf{W}_0^\top, \mathbf{W}_i^\top, \sqrt{\lambda}\mathbf{1}_{K\times 1})^\top$$
$$\mathbf{C} = (-\mathbf{I}_K, \mathbf{W}_0^\top, -\mathbf{W}_0^\top)^\top$$
$$\mathbf{D} = (-\mathbf{0}_{n\times K}, \mathbf{1}_{n\times n}^\top, -\mathbf{0}_{n\times n})^\top \tag{5}$$

Proof. It suffices to show the objective functions and constraints in Eq (3) and Eq (4) are correspondingly equivalent by constructing matrices $\mathbf{A}$, $\mathbf{B}$, $\mathbf{C}$, and $\mathbf{D}$.

When $\mathbf{W}_0$ and $\mathbf{W}_i$ are fixed, The last regularization, $\eta(\|\mathbf{W}_0\|_F^2 + \|\mathbf{W}_i\|_F^2)$, in Eq (3) is constant. Due to the nonnegative constraint on $\mathbf{H}$, $\sum_{j=1}^{n}\|\mathbf{H}(:,j)\|_1^2 = \|\mathbf{1}_{1\times K}H\|_2^2$. Then the objective function in Eq (3) can be converted to:

$$\|\mathbf{Y}_0 - \mathbf{W}_0\mathbf{H}\|_F^2 + \|\mathbf{Y}_i - \mathbf{W}_i\mathbf{H}\|_F^2 + \lambda\|\mathbf{e}_{1\times K}H\|_2^2 \tag{6}$$
$$= \|(\mathbf{Y}_0^\top, \mathbf{Y}_i^\top, \mathbf{0}_{n\times 1})^\top - (\mathbf{W}_0^\top, \mathbf{W}_i^\top, \sqrt{\lambda}\mathbf{1}_{K\times 1})^\top H\|_F^2$$
$$= \|\mathbf{A} - \mathbf{B}\mathbf{H}\|_F^2$$

It is easy to verify that the constraints in Eq (3) can be converted into:

$$(-\mathbf{I}_K, \mathbf{W}_0^\top, -\mathbf{W}_0^\top)^\top \mathbf{H} \leq (-\mathbf{0}_{n\times K}, \mathbf{1}_{n\times n}, -\mathbf{0}_{n\times n})^\top \tag{7}$$
$$= \mathbf{CH} \leq \mathbf{D}$$

which completes the proof.

For computing the component $\mathbf{W}_0$, we have the following theorem:

Theorem 2. *When components $\mathbf{W}_i$ and $\mathbf{H}$ are fixed, the formulation to optimize $\mathbf{W}_0$ in Eq (3) is equivalent to the following constrained minimization problem:*

$$\min_{\mathbf{W}_0} \|\mathbf{A} - \mathbf{B}\mathbf{W}_0^\top\|_F^2$$

$$\text{s.t.} \quad \mathbf{C}\mathbf{W}_0^\top \leq \mathbf{D} \tag{8}$$

where $\mathbf{A}$, $\mathbf{B}$, $\mathbf{C}$, *and* $\mathbf{D}$ *are defined as follows:*

$$\mathbf{A} = (\mathbf{Y}_0, \mathbf{0}_{K \times n})^\top$$
$$\mathbf{B} = (\mathbf{H}, \sqrt{\eta}\mathbf{I}_K)^\top$$
$$\mathbf{C} = (\mathbf{H}, -\mathbf{H})^\top$$
$$\mathbf{D} = (\mathbf{1}_{n \times n}, -\mathbf{0}_{n \times n})^\top \tag{9}$$

Proof. When $\mathbf{W}_i$ and $\mathbf{H}$ are fixed, $\|\mathbf{Y}_i - \mathbf{W}_i\mathbf{H}\|_F^2$, $\lambda \sum_{j=1}^n \|\mathbf{H}(:,j)\|_1^2$, and $\eta\|\mathbf{W}_i\|_F^2$ are constants. Then the objective function for $\mathbf{W}_0$ in Eq (3) is:

$$\|\mathbf{Y}_0 - \mathbf{W}_0\mathbf{H}\|_F^2 + \eta\|\mathbf{W}_0\|_F^2 \tag{10}$$
$$= \|(\mathbf{Y}_0, \mathbf{0}_{K \times n})^\top - (\mathbf{H}, \sqrt{\eta}\mathbf{I}_k)^\top \mathbf{W}_0^\top\|_F^2$$
$$= \|\mathbf{A} - \mathbf{B}\mathbf{W}_0^\top\|_F^2$$

The proof process for the equivalence of constraints is similar to that of Theorem 1.

From Theorem 2, we can see that given $\mathbf{H}$, the calculation of $\mathbf{W}_0$ is independent on $\mathbf{W}_i$ and $\mathbf{Y}_i$.

When $\mathbf{W}_0$ and $\mathbf{H}$ are fixed, since the three constraints in Eq (3) are independent of $\mathbf{W}_i$, the optimization problem for $\mathbf{W}_i$ is a typical least square problem:

$$\mathbf{W}_i = \mathbf{Y}_i\mathbf{H}^\top(\mathbf{H}\mathbf{H}^\top + \eta\mathbf{I}_K)^{-1} \tag{11}$$

Algorithm for Integrating Two Sources. Through Theorems 1 and 2, we notice that the optimization problems for computing $\mathbf{H}$ and $\mathbf{W}_0$ are equivalent in solving the following optimization problem:

$$\min_{\mathbf{X}} \|\mathbf{A} - \mathbf{B}\mathbf{X}\|_F^2$$

$$\text{s.t.} \quad \mathbf{C}\mathbf{X} \leq \mathbf{D} \tag{12}$$

this problem is indeed the collection of several linear constrained least square problems.

$$\min_{\mathbf{X}(:,j)} \|\mathbf{A}(:,j) - \mathbf{B}\mathbf{X}(:,j)\|_F^2$$

$$\text{s.t.} \quad \mathbf{C}\mathbf{X}(:,j) \leq \mathbf{D}(:,j) \tag{13}$$

In our implementation, we use the active-set method to solve this linear constrained least square problem and we assume that the function to solve Eq. (13) is named *iplsqlin*, has four arguments, and outputs the optimal x, i.e., $x = iplsqlin(\mathbf{a}, \mathbf{B}, \mathbf{C}, \mathbf{d})$. Algorithm 1 shows how to update $\mathbf{H}$. To get each column of $\mathbf{H}$, we have to solve a linear constrained least square problem.

Algorithm 1. Update-H

Input: The Link Matrix $\mathbf{Y}_0$, Source Matrix Y_i, the fixed components $\mathbf{W}_0$ and $\mathbf{W}_i$, and λ.

Output: H.

1: Construct $\mathbf{A}$, $\mathbf{B}$, $\mathbf{C}$, and $\mathbf{D}$ according to Eq. (5)
2: **for** $i = 1 \to n$ **do**
3:   $\mathbf{H}(:, i) \leftarrow iplsqlin(\mathbf{A}(:, i), \mathbf{B}, \mathbf{C}, \mathbf{D}(:, i))$
4: **end for**

Similar as the algorithm for updating $\mathbf{H}$, the algorithm for updating $\mathbf{W}_0$ is shown in Algorithm 2. The input of Algorithm 2 is independent on $\mathbf{Y}_i$ and $\mathbf{W}_i$.

Algorithm 2. Update-$\mathbf{W}_0$

Input: The Link Matrix $\mathbf{Y}_0$, the fixed components $\mathbf{H}$ and η.

Output: $\mathbf{W}_0$

1: Construct $\mathbf{A}$, $\mathbf{B}$, $\mathbf{C}$, and $\mathbf{D}$ according to Eq. (9)
2: **for** $i = 1 \to n$ **do**
3:   $\mathbf{W}_0^\top(:, i) \leftarrow iplsqlin(\mathbf{A}(:, i), \mathbf{B}, \mathbf{C}, \mathbf{D}(:, i))$
4: **end for**

Based on Update-$\mathbf{H}$ and Update-$\mathbf{W}_0$, we have Algorithm 3 to solve the problem in Eq. (3). Note that the solution of Eq. (3) is not unique. Given a solution of $\{\mathbf{W}_0, \mathbf{W}_i, \mathbf{H}\}$, $\{\mathbf{W}_0\mathbf{D}, \mathbf{W}_i\mathbf{D}, \mathbf{D}^{-1}\mathbf{H}\}$ is also the solution for Eq. (3), where $\mathbf{D}$ is a diagonal matrix with positive elements. We seek a unique solution by applying a normalization to each column of $\mathbf{H}$.

$$\mathbf{W}_0(j, k) = \mathbf{W}_0(j, k)\sqrt{\sum_j \mathbf{H}^2(j, k)}$$

$$\mathbf{W}_i(j, k) = \mathbf{W}_i(j, k)\sqrt{\sum_j \mathbf{H}^2(j, k)}$$

$$\mathbf{H}(j, k) = \frac{\mathbf{H}(j, k)}{\sqrt{\sum_j \mathbf{H}^2(j, k)}} \tag{14}$$

In Algorithm 3, after some initialization, we alternatively use Eq (11), Update-$\mathbf{H}$ and Update-$\mathbf{W}_0$ to update $\mathbf{W}_i$, $\mathbf{H}$ and $\mathbf{W}_0$ by fixing two of them. This alternative process will be iterated until convergence.

Illustration Based on a Toy Example. To further illustrate the advantages of the proposed framework for community detection, let us consider the example shown in Figure 1. There are two sources, i.e., link information and tag information. We run our two source method, i.e., Algorithm 3, to integrate link and tag information for community detection. The Affiliation Matrix $\mathbf{H}$ is shown as follows:

Algorithm 3. TwoSources

Input: The Link Matrix $\mathbf{Y}_0$, Source Matrix $\mathbf{Y}_i$, λ and η
Output: $\mathbf{H}$ and $\mathbf{W}_0$

1: Initialize $\mathbf{H}$, $\mathbf{W}_0$ and $\mathbf{W}_i$
2: **while** Not convergent **do**
3:   Update: $\mathbf{W}_i \leftarrow \mathbf{Y}_i \mathbf{H}^\top (\mathbf{H}\mathbf{H}^\top + \eta \mathbf{I}_K)^{-1}$
4:   Update: $\mathbf{H} \leftarrow$ Update-$\mathbf{H}(\mathbf{Y}_0, \mathbf{Y}_i, \mathbf{W}_0, \mathbf{W}_i, \lambda)$
5:   Update: $\mathbf{W}_0 \leftarrow$ Update-$\mathbf{W}_0(\mathbf{Y}_0, \mathbf{H}, \eta)$
6: **end while**
7: Normalize $\mathbf{H}$, $\mathbf{W}_0$ and $\mathbf{W}_i$ by Eq (14)

$$\mathbf{H} = \begin{pmatrix} 0\ 1\ 1\ 1\ .29\ .11\ 0\ 0\ 0\ \ \ 0 \\ 0\ 0\ 0\ 0\ .71\ .74\ 1\ 1\ 0\ .18 \\ 1\ 0\ 0\ 0\ \ 0\ .15\ 0\ 0\ 1\ .82 \end{pmatrix}$$

The first observation is that the solution is sparse and more than half of entities are exactly zeros. After normalization, $\mathbf{H}(i, j)$ is the probability of u_j belonging to c_i. We can see that the result is very consistent with the real memberships of users.

We also use $\mathbf{R}$ to reconstruct the original link matrix $\mathbf{Y}_0$. According to our framework, each column of $\mathbf{R}$, $\mathbf{R}(:, j)$, represents the strengths of relationships between u_j and other users. We examine $\mathbf{R}$ and find that the strengths of links $(4, 1)$ and $(4, 3)$ are much weaker than those of links $(4, 5)$, $(4, 6)$ and $(4, 7)$. We run Modularity Maximization on $\mathbf{R}$ and the result is shown in Figure 2, which is consistent with the ground truth, demonstrating the advantages of our proposed framework.

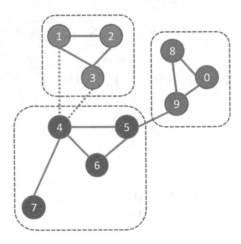

Fig. 2. Community Detection based on Two Sources

5   Integrating Multiple Sources

The development of the two-source solution paves the way for a multi-source solution. Given the link matrix $\mathbf{Y}_0$ and a set of m data sources $\mathcal{S} = \{\mathbf{Y}_1, \mathbf{Y}_2, ..., \mathbf{Y}_m\}$, the optimization problem for integrating multiple sources can be formalized as follows (mJointMF),

$$\min_{\mathbf{W}_0, \mathbf{W}_i, \mathbf{H}} \|\mathbf{Y}_0 - \mathbf{W}_0\mathbf{H}\|_F^2 + \sum_{i=1}^{m} \|\mathbf{Y}_i - \mathbf{W}_i\mathbf{H}\|_F^2$$

$$+ \lambda \sum_{j=1}^{n} \|\mathbf{H}(:,j)\|_1^2 + \eta \sum_{k=0}^{m} \|\mathbf{W}_k\|_F^2,$$

$$\text{s.t.} \quad \mathbf{W}_0\mathbf{H} \leq 1$$
$$\mathbf{W}_0\mathbf{H} \geq 0$$
$$\mathbf{H} \geq 0 \tag{15}$$

The following theorem shows the connection between the two source integration method and the multi-source integration method for community detection.

Theorem 3. *The optimization problem for multi-source integration is equivalent to the following minimization problem,*

$$\min_{\mathbf{W}_0, \mathbf{W}, \mathbf{H}} \|\mathbf{Y}_0 - \mathbf{W}_0\mathbf{H}\|_F^2 + \|\mathbf{C} - \mathbf{W}\mathbf{H}\|_F^2$$

$$+ \lambda \sum_{j=1}^{n} \|\mathbf{H}(:,j)\|_1^2 + \eta(\|\mathbf{W}_0\|_F^2 + \|\mathbf{W}\|_F^2),$$

$$\text{s.t.} \quad \mathbf{W}_0\mathbf{H} \leq 1$$
$$\mathbf{W}_0\mathbf{H} \geq 0$$
$$\mathbf{H} \geq 0 \tag{16}$$

where $\mathbf{W}$ *and* $\mathbf{C}$ *are defined as follows:*

$$\mathbf{C} = (\mathbf{Y}_1^\top, \mathbf{Y}_2^\top, \dots, \mathbf{Y}_m^\top)^\top$$
$$\mathbf{W} = (\mathbf{W}_1^\top, \mathbf{W}_2^\top, \dots, \mathbf{W}_m^\top)^\top$$

Proof. Comparing Eq (15) with Eq (16), there are two differences. Therefore, it suffices to show that they are correspondingly equivalent. The following formulation suggests that the first difference is equivalent.

$$\sum_{i=1}^{m} \|\mathbf{Y}_i - \mathbf{W}_i\mathbf{H}\|_F^2$$

$$= \|(\mathbf{Y}_1^\top, \mathbf{Y}_2^\top, \dots, \mathbf{Y}_m^\top)^\top - (\mathbf{W}_1^\top, \mathbf{W}_2^\top, \dots, \mathbf{W}_m^\top)^\top \mathbf{H}\|_F^2$$

$$= \|\mathbf{C} - \mathbf{W}\mathbf{H}\|_F^2 \tag{17}$$

The equivalence of the second difference is shown below:

$$\eta \sum_{k=0}^{m} \|\mathbf{W}_k\|_F^2$$
$$= \eta(\|\mathbf{W}_0\|_F^2 + \|(\mathbf{W}_1^\top, \mathbf{W}_2^\top, \ldots, \mathbf{W}_m^\top)^\top\|_F^2)$$
$$= \eta(\|\mathbf{W}_0\|_F^2 + \|\mathbf{W}\|_F^2) \tag{18}$$

which completes the proof.

Theorem 3 implies that the optimization problem for integrating multiple sources is equivalent to that for integrating two sources. The significance of this theorem is twofold: first, it provides a way to solve the multiple sources integration problem using two sources integration, which is shown in Algorithm 4; and second, it provides an intuitive explanation for how the data of multiple sources are integrated: *sources in $\mathcal{S}$ firstly stack together and then integrate with the link source*.

Algorithm 4. MultipleSource

Input: The Link Matrix $\mathbf{Y}_0$, the source set $\mathcal{S}$, λ and η.
Output: $\mathbf{H}$ and $\mathbf{W}_0$
  1: Construct $\mathbf{C}$ according to Eq. (9)
  2: Set $(\mathbf{H}, \mathbf{W}_0) = \text{TwoSources}(\mathbf{Y}_0, \mathbf{C}, \lambda, \eta)$
  3: Normalize $\mathbf{H}$ and $\mathbf{W}_0$ by Eq (14)

---

6   Experimental Evaluation

To verify the effectiveness of our proposed method, we conduct experiments on both synthetic data and real-world social media data.

6.1   Synthetic Data

Since the ground truth is usually unavailable for real-world social media data, we resort to synthetic data to show if the proposed framework can achieve the design goals. The synthetic data consists of two types of information: link information and tag information. Parameters in generating the synthetic data include the number of users, n, the number of tags, t, the number of communities, K, link (tag) density within and between communities, ρ_w, ρ_b, and the ratio of noise links (tags) ρ_n. To generate the ground truth, users and tags are split evenly into each community, and according to link (tag) density within communities ρ_w, we randomly generate links between users (users and tags) in the same community. While relying on link (tag) density between communities ρ_b, we randomly create links between users (users and tags) from different communities.

   To simulate noise and complementary information in sources, we design the following procedure,

- Randomly assign communities into two groups with equal size, i.e., g_1 and g_2. Let u_1 and t_1 be the set of users and tags in g_1, respectively, while u_2 and t_2 are the set of users and tags in g_2.
- Randomly add links between u_1 according to the noise ratio ρ_n, for link information. For tag information, we randomly add links for u_2.

Through the above process, with link information for u_2 being fixed, we add noisy tags to u_2, and with tag information for u_1 being fixed, noisy links are added to u_1. Therefore, link and tagging information generated above are noisy but complementary with each other.

In this experiment, we generate a set of datasets with parameters: $n = 1000$, $t = 1000$, $k = 20$, $\rho_w = 0.8$, $\rho_b = 0.1$ and varying ρ_n from 0 to 1 with step 0.1. Five baseline methods are used: LDA-Link(LL) [4], PCL-Link(PL) [25], EdgeCluster(EC) [19] and Modularity Maximization(Modu) [12] with only link information; and Tag-CoClustering(TC) [22] using only tagging information. All parameters in comparing methods are determined by cross validation. Normalized mutual information is adopted to evaluate the community quality. The average NMI performance w.r.t the noise ratio ρ_n are shown in Figure 3

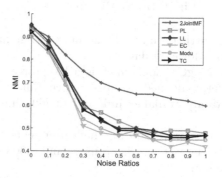

Fig. 3. NMI Performance w.r.t Noise Ratios

The first observation is that with the noise ratios increasing, all performance reduces dramatically, especially for link-based algorithms. This supports our assumption that noise links or weak links in online social networks can make link-based algorithms ineffective. By integrating two sources, our algorithm consistently outperforms algorithms with a single source.

6.2 Social Media Data

We use data from real-world social media websites, i.e., BlogCatalog[3] and Flickr[4]. The first two datasets are obtained from[19]. We crawled the third dataset to

---

[3] http://www.blogcatalog.com
[4] http://www.flickr.com/

Table 1. Statistics of the Datasets

	BC	Flickr	BC-MS
# of Users	8,797	8,465	6,069
# of Links	290,059	195,847	523,642
# of Sources	2	2	4
Ave Degree	66	46	173
Density	0.0075	0.0055	0.028
Clustering Coefficient	0.46	0.13	0.39

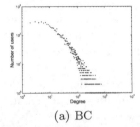
(a) BC

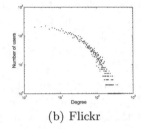
(b) Flickr

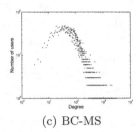
(c) BC-MS

Fig. 4. Degree Distributions

include four sources for further study. The number of communities, K, is determined by cross validation for each dataset.

BC is crawled from BlogCatalog, which is a blog directory where users can register their blogs under predefined categories. It contains 8,797 users and 7,418 tags. Two types of data are available: link and tagging information, and $K = 1,000$.

Flickr is an image sharing website in which users can specify tags for each image they upload. The dataset has 8,465 users and 7,303 tags with both link and tagging information, and $K = 500$.

BC-MS is collected from BlogCatalog with two additional sources besides link and tagging data: commenting and reading. It has 6,069 users and 5,161 tags. The four sources are S1 (linking), S2 (tagging), S3 (commenting), and S4 (reading), and $K = 500$.

Some statistics of the datasets are shown in Table 1. We also compute the degree for each user. The distributions are shown in Figure 4, suggesting a power law distribution that is typical in social networks.

Since there is no ground truth about the online communities and the discovered communities are overlapping, we cannot compute the traditional metrics such as NMI and Modularity. Thus, we evaluate the quality of identified communities indirectly, which has been adopted by [22]. The basic assumption is that users belonging to the same communities should exhibit similar behaviors. Treating cluster memberships as features, randomly selecting a certain fraction of instances as training data and the rest as testing data, the evaluation is turned into a classification problem. We obtain the labels for each user from the social networking websites. In our work, the users' interests are treated as labels for each user. Linear SVM is adopted in our experiments since it scales well to large data sets. The training data size varies from 10% to 90% of the whole data. The experiments are repeated 10 times by shuffling data each time. Average Micro-F1 and Macro-F1 measures are reported.

Integrating Two Sources. Cross validation is employed to determine the value of the regularization parameters, i.e., λ and η. We set λ to 0.05 in Flickr dataset and λ to 0.1 in both BC and BC-MS. Set η to 0.05 in all datasets. The iteration is stopped until the difference of the objective function between two consecutive steps is smaller than 1e-6. We focus on the two-source integration as in Eq. (3), which integrates link and another source (tagging, commenting, or reading) and compare it with single-source methods. PL, LL, EC and Modu work with the link matrix, and TC applies co-clustering to the user-tag data.

Tables 2 and 3 show the prediction performance on BC and Flickr, respectively. The first observation is that the prediction performance improves as when more training data is used. PL, LL, EC and TC show comparable performance for both datasets. The proposed integrative method using both links and tagging information outperforms the single-source methods significantly. Compared to the best performance of baseline methods, on average, we achieve 17.2% and 31.5% improvement with respect to Micro-F1 in BC and Frickr, respectively. We obtain similar improvement in terms of Macro-F1. This directly supports that integrating different types of data in social media significantly improves the performance of community detection.

In addition, we report in Table 4 the performance of combining links with other data sources on BC-MS, such as S3 (commenting) and S4 (reading). Integrating an additional data source leads to much better performance. We also observe that different sources make uneven contributions to community quality: tagging information being the most, followed by commenting, and then reading. This implies that improvement might rely on the quality of sources.

Comparative Study of Integrative Methods. In this section, we study performance of different data integration methods on BC-MS. We compare our

Table 2. Performance on BC Dataset

Proportion of Labeled Nodes		10%	20%	30%	40%	50%	60%	70%	80%	90%
Micro-F1(%)	2JointMF	**44.53**	**46.35**	**50.11**	**50.41**	**52.05**	**52.12**	**52.99**	**53.03**	**53.12**
	PL	28.94	28.85	30.85	31.20	32.25	33.10	33.11	33.42	33.60
	LL	26.61	26.24	26.57	26.73	27.74	26.63	27.50	27.38	27.99
	EC	24.85	25.55	26.27	25.18	25.28	24.80	24.11	23.94	22.22
	Modu	16.46	20.38	19.46	21.20	23.13	21.51	22.68	22.39	22.66
	TC	38.45	37.75	40.53	38.84	41.92	41.30	43.77	43.15	44.88
Macro-F1(%)	2JointMF	**29.01**	**31.12**	**34.76**	**35.54**	**36.99**	**37.59**	**38.02**	**39.11**	**39.39**
	PL	15.38	16.30	17.30	18.18	18.38	18.72	18.71	17.61	18.13
	LL	15.49	15.32	16.25	15.94	15.85	16.08	16.11	15.88	16.74
	EC	14.24	15.16	16.43	15.75	15.96	16.08	15.42	15.78	14.99
	Modu	9.32	9.34	10.61	11.39	10.53	11.01	11.01	9.69	11.66
	TC	28.85	26.83	27.68	28.52	28.18	29.69	28.60	30.16	29.96

Table 3. Performance on Flickr Dataset

Proportion of Labeled Nodes		10%	20%	30%	40%	50%	60%	70%	80%	90%
Micro-F1(%)	2JointMF	**55.99**	**54.31**	**55.57**	**54.76**	**54.51**	**54.78**	**54.99**	**55.57**	**57.02**
	PL	42.03	44.53	44.72	45.22	46.68	47.68	47.90	48.43	49.27
	LL	40.80	41.17	42.49	42.55	43.13	44.16	45.69	45.88	46.51
	EC	39.62	39.93	40.93	41.12	41.79	41.75	42.06	42.57	43.44
	Modu	29.72	31.69	32.06	32.28	33.35	33.04	34.25	34.20	34.82
	TC	37.42	37.80	37.90	38.35	39.08	39.22	39.35	39.99	40.12
Macro-F1(%)	2JointMF	**30.62**	**30.81**	**31.13**	**31.49**	**32.04**	**32.12**	**31.99**	**32.11**	**32.42**
	PL	20.16	20.25	20.46	20.50	20.10	19.95	20.31	20.29	20.40
	LL	19.83	20.19	20.55	20.58	20.81	21.08	21.43	21.45	22.09
	EC	20.83	20.66	21.03	20.74	20.86	20.51	20.90	20.87	21.11
	Modu	15.35	13.25	13.45	13.37	13.10	13.29	13.78	13.92	14.14
	TC	20.65	20.49	21.03	20.90	20.80	20.68	21.06	21.28	21.35

Table 4. Performance on BC-MS Dataset

Proportion of Labeled Nodes		10%	20%	30%	40%	50%	60%	70%	80%	90%
Micro-F1(%)	2JointMF(S1+S2)	**29.95**	**32.87**	**33.47**	**34.23**	**34.53**	**34.80**	**35.04**	**34.56**	**34.91**
	2JointMF(S1+S3)	27.30	29.39	29.53	31.59	31.80	31.99	32.17	31.05	32.21
	2JointMF(S1+S4)	25.06	26.10	27.09	27.47	27.83	28.33	28.74	28.31	27.04
	PL	13.94	14.80	14.73	14.63	14.92	15.62	16.44	16.89	15.85
	LL	16.27	16.98	17.97	18.52	18.46	18.55	18.92	19.33	18.38
	EC	14.10	14.40	14.96	15.58	16.13	16.47	16.72	16.45	15.78
	Modu	9.25	12.81	12.85	14.41	12.62	13.22	13.49	14.15	13.69
	TC	14.79	14.69	15.32	15.95	15.63	15.24	16.31	16.91	16.50
Macro-F1(%)	2JointMF(S1+S2)	**9.66**	**11.14**	**12.75**	**13.01**	**13.08**	**13.11**	**13.07**	**12.75**	**13.05**
	2JointMF(S1+S3)	8.15	8.75	8.83	8.99	8.99	9.18	8.91	9.80	10.17
	2JointMF(S1+S4)	7.51	8.33	8.51	9.12	9.08	9.44	9.60	9.55	9.04
	PL	3.62	3.60	3.54	3.81	4.50	4.52	4.97	4.86	5.43
	LL	5.59	5.69	6.18	6.36	6.24	6.27	6.65	6.35	6.71
	EC	3.04	3.52	3.96	4.19	4.46	4.67	4.71	4.84	4.53
	Modu	1.92	2.57	2.81	3.22	2.71	2.71	3.13	3.10	2.95
	TC	4.06	4.11	4.58	4.85	4.91	5.04	5.12	5.00	5.25

multi-source method with three integrative baseline methods. PMM [20] first extracts the top eigenvectors of multiple data sources and combines them into a principal matrix, then obtain an overlapping clustering. Similarly, Canonical Correlation Analysis (CCA) can be used to find a transformation matrix for each source matrix such that the pairwise correlations between the projected matrices are maximized [3], and overlapping communities are then extracted. Cluster-ensemble [15] is adopted in this work to first compute the affiliation matrices for data sources, then combine them to find a consensus clustering. Note that these baseline integration methods have two stages: 1) integrating multi-source; 2) performing traditional community detection methods. In this experiment, since the input matrix can be negative, EdgeCluster is adopted as the basic community detection algorithm. However, our method performs multi-source integration and community detection simultaneously. All sources on BC-MS (linking, tagging, commenting, and reading) are integrated. The results are presented in Figures 5(a) and 5(b), respectively.

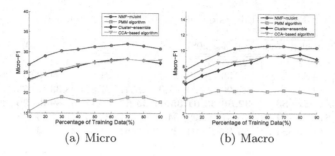

(a) Micro   (b) Macro

Fig. 5. Comparisons of Different Integrating Schemes

mJointMF gains 14.4% and 14.8% improvement of relative ratio compared with CCA-based method and Cluster-ensemble in terms of Micro, respectively. And it improves with relative ratios 15.7% and 20.5% compared with CCA-based method and Cluster-ensemble w.r.t Macro, respectively. In both cases, CCA-based and Cluster-ensemble have similar performance, however, PMM does not fare well.

Different Returns of Various Data Sources. In this subsection, we try to investigate whether performance always improves as the number of sources increases. In earlier experiments, we observe that integrating an additional source with link data consistently improves performance over using only link information. We systematically examine performance by adding S2 (tagging), S3 (commenting), and S4 (reading) to S1 (linking).

As seen in Table 5, the benefit of having more data sources is not linearly associated with performance. Peak performance is achieved when integrating linking (S1), tagging (S2), and commenting (S3) in most cases. In some cases, adding another source can also worsen performance. Theorem 3 suggests that integration multiple source is divided into two phrases: 1) stacking other sources together; and 2) integrating it with link information. When more sources are added, the dimension will be increased significantly which will make the algorithm ineffective because of the curse of dimensionality; more noise may be introduced when more data sources are integrated; and redundant information may also exist in the sense that one source offers no new information due to the availability of other sources.

Validating Relationship Strengths for Community Detection. In this section, we study how useful the estimated relationship strengths for link-based community detection algorithms. That is to say, we want to investigate if the link strengths estimated by our framework can help improve the performance of link-based algorithms. Four representative link-based algorithms are adopted in this experiment: PL, LL, EC, and Modu. The average Micro and Macro performance on 10 runs with 50% training dataset in BC and BC-MS datasets are shown in Figure 6 and Figure 7 respectively since similar results can be observed with other settings.

Table 5. Effects of Integrating Different Sources

Proportion of Labeled Nodes		10%	20%	30%	40%	50%	60%	70%	80%	90%
Micro-F1(%)	S1+S2	29.95	32.87	33.47	34.23	34.53	34.80	35.04	34.56	34.91
	S1+S3	27.30	29.39	29.53	31.59	31.80	31.99	32.17	31.05	32.21
	S1+S4	25.06	26.10	27.09	27.47	27.83	28.33	28.74	28.31	27.04
	S1+S2+S3	**32.56**	**33.61**	**35.01**	**35.67**	36.12	35.89	**36.99**	36.24	**36.90**
	S1+S2+S4	31.41	32.99	34.66	35.21	**36.54**	**36.11**	36.47	**36.41**	36.60
	S1+S3+S4	27.02	28.04	28.53	29.62	31.03	31.26	31.88	31.18	29.84
	S1+S2+S3+S4	26.90	28.98	30.21	30.61	31.27	31.53	31.91	31.40	30.76
Macro-F1(%)	S1+S2	9.66	11.14	12.75	13.01	13.08	13.11	13.07	12.75	13.05
	S1+S3	8.15	8.75	8.83	8.99	8.99	9.18	8.91	9.80	10.17
	S1+S4	7.51	8.33	8.51	9.12	9.08	9.44	9.60	9.55	9.04
	S1+S2+S3	**11.24**	11.60	**13.02**	**14.51**	**14.99**	**15.01**	**14.92**	**15.04**	**15.27**
	S1+S2+S4	10.96	**12.41**	12.99	13.19	13.48	13.99	14.38	14.66	14.84
	S1+S3+S4	8.54	8.51	8.47	8.68	9.17	8.91	9.24	9.80	8.65
	S1+S2+S3+S4	7.54	8.67	9.61	10.28	10.47	10.65	10.60	10.33	10.38

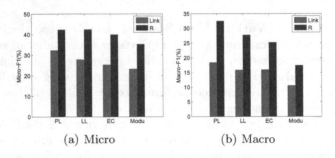

(a) Micro   (b) Macro

Fig. 6. Comparisons of Performance in BC. Note that "Link" denotes performance on original link information and "R" represents performance on link information with estimated strengths

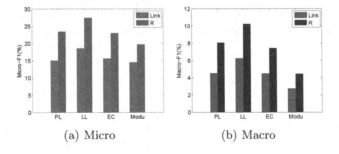

(a) Micro   (b) Macro

Fig. 7. Comparisons of Performance in BC-MS. Note that "Link" denotes performance on original link information and "R" represents performance on link information with estimated strengths

The performance of all four link-based algorithms is significantly improved. For example, on average, PL gains 34.4% and 66.7% improvement of relative ratio with respect to Micro performance in BC and BC-MS, respectively. And it improves with relative ratio 83.3% and 71.1% w.r.t Macro performance in BC and BC-MS, respectively. We have similar observations for LL, EC and Modu as well. These results indicate that the relationship strengths estimated by our framework can significantly improve the performance of link-based community detection algorithms.

7   Conclusions

In this work, we study how to utilize social media data of different types for detecting communities. We propose an optimization framework to integrate multiple sources for community detection and estimating link strengths. Experimental results show promising findings: (1) integrating multiple data sources helps improve the performance of community detection; (2) different sources contribute unevenly to performance improvement of community detection; (3) having more data sources does not necessarily bring about better performance; and (4) the relationship strengths estimated by our framework can significantly improve the performance of link-based community detection algorithms.

This study also suggests some interesting problems for further exploration. Experimental results reveal that performance improvement might rely on the quality of sources. In order to find the relevant sources, we need efficient ways of studying the relationships between different sources as it is impractical to enumerate all sources to determine relevant sources even when the number of sources is moderately large. Exploring additional sources of social media data is another promising direction, e.g., incomplete user profiles, short and unconventional text like tweets may also be helpful.

Acknowledgments. The work is, in part, supported by ARO (#025071) and NSF (#0812551).

References

1. Backstrom, L., Huttenlocher, D., Kleinberg, J., Lan, X.: Group formation in large social networks: membership, growth, and evolution. In: KDD, pp. 44–54. ACM (2006)
2. Blei, D.M., Ng, A.Y., Jordan, M.I.: Latent dirichlet allocation. The Journal of Machine Learning Research 3, 993–1022 (2003)
3. Chaudhuri, K., Kakade, S.M., Livescu, K., Sridharan, K.: Multi-view clustering via canonical correlation analysis. In: ICML (2009)
4. Erosheva, E., Fienberg, S., Lafferty, J.: Mixed-membership models of scientific publications. Proceedings of the National Academy of Sciences of the United States of America 101(suppl. 1), 5220 (2004)
5. Evans, T., Lambiotte, R.: Line graphs, link partitions, and overlapping communities. Physical Review E 80(1), 16105 (2009)

6. Lin, Y.-R., Sun, J., Castro, P., Konuru, R., Sundaram, H., Kelliher, A.: Metafac: community discovery via relational hypergraph factorization. In: KDD, pp. 527–536. ACM (2009)
7. Liu, Y., Niculescu-Mizil, A., Gryc, W.: Topic-link lda: Joint models of topic and author community. In: ICML 2009 (2009)
8. Luxburg, U.: A tutorial on spectral clustering. Statistics and Computing 17(4), 395–416 (2007)
9. McPherson, M., Lovin, L.S., Cook, J.M.: Birds of a feather: Homophily in social networks. Annual Review of Sociology 27(1), 415–444 (2001)
10. Newman, M.E.: Finding community structure in networks using the eigenvectors of matrices. Physical Review E 74(3), 36104 (2006)
11. Newman, M.E., Leicht, E.: Mixture models and exploratory analysis in networks. Proceedings of the National Academy of Sciences 104(23), 9564 (2007)
12. Newman, M.E.J., Girvan, M.: Finding and evaluating community structure in networks. Physical Review E 69(2), 26113 (2004)
13. Palla, G., Dernyi, I., Farkas, I., Vicsek, T.: Uncovering the overlapping community structure of complex networks in nature and society. Nature 435(7043), 814–818 (2005)
14. Scellato, S., Mascolo, C., Musolesi, M., Latora, V.: Distance matters: Geo-social metrics for online social networks. In: WOSN 2010 (2010)
15. Strehl, A., Ghosh, J.: Cluster ensembles – a knowledge reuse framework for combining multiple partitions. Journal of Machine Learning Research 3, 583–617 (2002)
16. Tang, J., Gao, H., Liu, H.: mtrust: discerning multi-faceted trust in a connected world. In: WSDM, pp. 93–102. ACM (2012)
17. Tang, J., Liu, H.: Feature selection with linked data in social media. In: SDM (2012)
18. Tang, J., Liu, H.: Unsupervised feature selection for linked social media data. In: KDD (2012)
19. Tang, L., Liu, H.: Scalable learning of collective behavior based on sparse social dimensions. In: CIKM, pp. 1107–1116. ACM (2009)
20. Tang, L., Wang, X., Liu, H.: Uncovering groups via heterogeneous interaction analysis. In: ICDM, Miami, FL, USA, December 6-9 (2009)
21. Tibshirani, R.: Regression shrinkage and selection via the lasso. Journal of the Royal Statistical Society. Series B 58(1), 267–288 (1996)
22. Wang, X., Tang, L., Gao, H., Liu, H.: Discovering overlapping groups in social media. In: ICDM, Sydney, Australia, December 14 - 17 (2010)
23. White, S., Smyth, P.: A spectral clustering approach to finding communities in graphs. In: SDM, p. 274. Society for Industrial Mathematics (2005)
24. Xiang, R., Neville, J., Rogati, M.: Modeling relationship strength in online social networks. In: WWW, pp. 981–990. ACM (2010)
25. Yang, T., Jin, R., Chi, Y., Zhu, S.: Combining link and content for community detection: a discriminative approach. In: KDD, pp. 927–936. ACM (2009)

Face-to-Face Contacts at a Conference:
Dynamics of Communities and Roles

Martin Atzmueller[1], Stephan Doerfel[1], Andreas Hotho[2],
Folke Mitzlaff[1], and Gerd Stumme[1]

[1] University of Kassel, Knowledge and Data Engineering Group,
Wilhelmshöher Allee 73, 34121 Kassel, Germany
[2] University of Würzburg, Data Mining and Information Retrieval Group,
Am Hubland, 97074 Würzburg, Germany
{atzmueller,doerfel,mitzlaff,stumme}@cs.uni-kassel.de,
hotho@informatik.uni-wuerzburg.de

Abstract. This paper focuses on the community analysis of conference participants using their face-to-face contacts, visited talks, and tracks in a social and
ubiquitous conferencing scenario. We consider human face-to-face contacts and
perform a dynamic analysis of the number of contacts and their lengths. On
these dimensions, we specifically investigate user-interaction and community
structure according to different special interest groups during a conference.
Additionally, using the community information, we examine different roles and
their characteristic elements.

The analysis is grounded using real-world conference data capturing community information about participants and their face-to-face contacts. The analysis
results indicate, that the face-to-face contacts show inherent community structure
grounded using the special interest groups. Furthermore, we provide individual
and community-level properties, traces of different behavioral patterns, and
characteristic (role) profiles.

1 Introduction

During a conference, social contacts form an essential part of the experience of the
participants. Commonly, the term "networking" is used for describing the inherent
processes in such interactions. Typically, there are different (implicit and explicit) communities present at a conference, defined according to interests or membership to certain
tracks or special interest groups. In order to enhance the conferencing experience,
e.g., for obtaining interesting recommendations for papers and contacts, knowledge
discovery techniques can often be applied. During a conference, ubiquitous computing
approaches, e.g., based on RFID-tokens, provide dynamic adaptation options.

In this paper, we focus on the analysis of social data and contact patterns of
conference participants: We consider communities of participants and their visited talks
and we analyze their face-to-face contacts in these contexts. We examine different
explicit and implicit roles of the participants, validate the community structures, and
analyze various structural properties of the contact graph.

Our contribution is three-fold: We present an in-depth analysis of the social relations
and behavioral patterns at the conference, analyze different community structures,

M. Atzmueller et al. (Eds.): MSM/MUSE 2011, LNAI 7472, pp. 21–39, 2012.

and identify characteristics of special roles and groups. We focus on profiles of the participants and their dynamics considering face-to-face contacts, and a set of different community structures. For these, we use grounded information considering special interest groups, as given by the participants during registration in comparison to the emerging communities at the conference. In this way, we investigate how the established community structures emerge in the dynamic conference context. Finally, we describe and analyze different roles and groups, e.g., organizers and different subcommunities at a conference, in order to identify distinguishing characteristics.

The rest of the paper is structured as follows: Section 2 discusses social applications for conferences, and issues of privacy and trust. After that, Section 3 considers related work. Next, Section 4 provides the grounding of our approach presenting an in-depth analysis and evaluation of real-world conference data. Finally, Section 5 concludes the paper with a summary and interesting directions for future research.

2   Social Conferencing

During a conference, participants encounter different steps and phases: Preparation (before the conference), during the actual conference, and activities after the conference. Appropriate talks and sessions of interest need to be selected; talks and discussions need to be memorized. Additionally, social contacts during a conference are often essential, e.g., for networking, and are often revisited after a conference, as are the visited talks. All of these steps are supported by the system CONFERATOR[1]: It is in joint development by the School of Information Sciences, University of Pittsburgh (conference management component, as a refinement of the Conference Navigator [1]) and the Knowledge and Data Engineering group at the university of Kassel (social and ubiquitous PEERRADAR component).

A first prototype of CONFERATOR [2], developed by the Knowledge and Data Engineering group, has been successfully deployed at the LWA 2010 conference at the University of Kassel in October 2010. The system is based on the UBICON framework[2]: The PEERRADAR is applied for managing social and ubiquitous/real contacts, supported by embedding social networks such as Facebook, XING, and LinkedIn, while the TALKRADAR enables the personalization of the conference program. Furthermore, active RFID proximity-tags, cf., [3], provide for detecting the location and contacts between conference participants.

In CONFERATOR, privacy is a crucial issue: A variety of user data is collected. Therefore, appropriate steps for their secure storage and access were implemented. CONFERATOR implements privacy measures using a refined trust system: It features several privacy levels (private, trusted, public) for organizing access, e.g., to location or contact information. In the analysis, we aim at providing implicit k-anonymity in the presentation and discussion: We provide results at the level of special interest groups, or (for more detailed results) at groups containing at least four participants.

---

[1] http://www.conferator.org
[2] http://www.ubicon.eu

3 Related Work

Regarding the tracking and analysis of conference participants, there have been several approaches, using RFID-tokens or Bluetooth-enabled devices. Hui et al. [4] describe an application using Bluetooth-based modules for collecting mobility patterns of conference participants. Furthermore, Eagle and Pentland [5] present an approach for collecting proximity and location information using Bluetooth-enabled mobile phones.

One of the first experiments using RFID tags to track the position of persons on room basis was conducted by Meriac et al. (cf., [6]) in the Jewish Museum Berlin in 2007. Cattuto et al. [7] added proximity sensing in the Sociopatterns project[3]. Isella et al. [8] conducted further experiments on a variety of contact networks obtained via RFID technology. Alani and colleagues, e.g., [3], also added contact information from social online networks in the live social semantics experiments. In [9], the authors analyze social dynamics of conferences focussing on the social activity of conference participants in those experiments. They analyze, for example, their activity in social web platforms like Facebook, Twitter and other social media together with status and their research seniority. These experiments have also extended their focus from conferences to schools [10] and hospitals [11].

Our work uses the same technical basis (RFID tokens with proximity sensing) as the Sociopatterns project, which allows us to verify their very interesting results independently. Furthermore, in this paper we significantly extend the analysis: We are able to use further techniques in order to characterize different roles, communities and participant relations. Concerning this, we especially focus on the community and role dynamics which are enabled by given gold-standard community structures according to a set of special interest groups. The conference navigator by Brusilovsky [1] allows researchers attending a conference to organize the conference schedule and provides a lot of interaction capabilities. However, it is not connected to the real live activity of the user during the conference. In our application, we measured face-to-face conctacts, increased the precision of the localization component compared to previous RFID-based approaches, and linked together tag information and the schedule of a workshop week. Furthermore, we implemented a light-weight integration with BibSonomy and other social systems used by participants. This is the basis for new insights into the behavior of all participants.

Thus, in comparison to the approaches mentioned above, we are able to perform a much more comprehensive evaluation of the patterns acquired in a conference setting, since our data provides a stable ground truth for communities (the special interest groups). These provide a grounding not only for the verification of the structural properties of the contact patterns, but also for the roles and communities.

Identifying different "roles" of nodes and finding so called "key actors" has attracted a lot of attention, ranging from different measures of centrality (cf., [12]) to the exploration of topological graph properties [13] or structural neighborhood similarities [14]. We focus on a metric that measures, how much a node connects

---

[3] http://www.sociopatterns.org

different communities, cf., [15]. It can be based on initially given community structures or on a probabilistic model.

4   Grounding

In this section, we present an analysis of the collected conferencing data. After introducing some preliminaries, we first discuss a grounding of the communities given through the assignment of participants to special interest groups. After that, we consider roles concerning academic status (Prof., PostDoc, PhD-Student and Student) and roles concerning the conference organization (organizers and regular participants).

4.1   Preliminaries

In the following section, we briefly introduce basic notions, terms and measures used throughout this paper. We presume familiarity with the concepts of *directed* and *undirected* Graphs $G = (V, E)$ for a finite set V of nodes with edges $(u, v) \in V \times V$ and $\{u, v\} \subseteq V$ respectively. In a *weighted* graph, each edge is associated with a corresponding edge weight, typically given by a mapping from E to $\mathbb{R}$. We freely also use the term *network* as a synonym for a graph. For more details, we refer to standard literature, e.g., [16,17].

In the context of social network analysis, a *community* within a graph is defined as a group of nodes such that group members are densely connected among each other but sparsely connected to nodes outside the community [18]. Community structure was observed in several online social networks [19,20] and is sometimes also called "virtual community" [21].

For a given undirected graph $G = (V, E)$ and a community $C \subseteq V$ we use the following notation: $n := |V|$, $m := |E|$, $n_C := |C|$, $m_C := |\{\{u, v\} \in E : u, v \in C\}|$ – the number of *intra-edges* of C, and $\bar{m}_C := |\{\{u, v\} \in E : |\{u, v\} \cap C| = 1\}|$ – the number of *inter-edges* of C.

For formalizing and assessing community structure in networks, this work focuses on two prominent measures: The *modularity* [18] and the *segregation index* [22]. The modularity measure is based on a comparison of the number of edges within a community to the expected such number given a null-model (i.e., a randomized graph). Thus, the modularity of a (disjoint) community clustering is defined to be the fraction of the edges that fall within the given clusters minus the expected such fraction if edges were distributed at random. This can be formalized as follows: The modularity MOD of a graph clustering is given by

$$\text{MOD} = \frac{1}{2m} \sum_{i,j} \left(A_{i,j} - \frac{k_i k_j}{2m} \right) \delta(C_i, C_j) \,,$$

where A is the adjacency matrix, C_i and C_j are the clusters containing the nodes i and j respectively, k_i and k_j denote the degree of i and j. Further, $\delta(C_i, C_j)$ is the *Kronecker delta* symbol that equals 1 if $C_i = C_j$, and 0 otherwise.

A straightforward generalization of the above formula to a modularity measure wMOD in weighted networks [23] considers A_{ij} to be the weight of the edge between nodes i and nodes j, and replaces the degree k_i of a node i by its strength $s(i) = \sum_j A_{ij}$, i. e., the sum of the weights of the attached edges. For overlapping communities, the modularity can also be generalized as described in [24].

The *segregation index* SIDX [22] compares the number of expected inter-edges to the number of observed inter-edges, normalized by the expectation:

$$\text{SIDX}(C) = \frac{E(\bar{m}_C) - \bar{m}_C}{E(\bar{m}_C)} = 1 - \frac{\bar{m}_C n(n-1)}{2mn_C(n-n_C)}.$$

Thus, the segregation index explicitly includes the size n of the graph. By averaging the segregation over all clusters one obtains the segregation of a cluster model.

4.2 Available Data

For capturing social interactions, RFID proximity tags of the Sociopatterns project were handed out to the participants of the conference LWA 2010. 70 out of 100 participants volunteered to wear an RFID tag which (approximately) detected mutual face-to-face sightings among participants with a minimum proximity of about one meter. Each such sighting with a minimum length of 20 seconds was considered a contact which ended when the corresponding tags did not detect an according sighting for more than 60 seconds.

Using the recorded contact data, we generated undirected networks LWA$[\geq i]_*$, LWA$[\geq i]_\Sigma$, and LWA$[\geq i]_\#$. An edge $\{u, v\}$ is created, iff a contact with a duration of at least i *minutes* among participants u and v was detected ($i = 1, \ldots, 15$). For $i \geq 5$[minutes], for example, we can filter out "small talk" conversations. In LWA$[\geq i]_\#$ each edge is weighted with the *number* of according contacts, in LWA$[\geq i]_\Sigma$ it is weighted with the *sum* of all according contact durations, whereas LWA$[\geq i]_*$ is unweighted.

Table 1 contains some statistics for LWA$[\geq i]_*$, $i = 0, 5, 10$. The diameters and average path lengths coincide with those given in [8] for the Hypertext Conference 2009 (HT09). For LWA$[\geq 0]_*$, Figure 1 shows the degree and contact length distributions. The latter exhibits characteristics that are comparable with those given for HT09, whereas the degree distributions differ by exhibiting two peaks – one around 10 and one around 20 – in contrast to only one peak around 15 for HT09. We hypothesize that this deviation is due to a more pronounced influence of the conference organizers at LWA 2010 in relation to the total number of participants (approx. 15% of the participants in LWA$[\geq 0]_*$ were organizers). This hypothesis is supported by removing all organizers from LWA$[\geq 0]_*$ and recalculating the degree distribution, yielding a single peak in the interval $[15, 20]$.

Other statistics (e. g., strength distribution) also suggest evidence for structural similarities among HT09 and LWA 2010. Therefore, we conclude, that LWA 2010 was a typical technical conference setup, and results obtained at the LWA 2010 are expected to hold in other conference scenarios with similar size, too.

Table 1. High level statistics for different networks: Number of nodes and edges, Average degree, Average path length APL, diameter d, clustering coefficient C, number and size of the largest weakly connected component #CC and $|CC|_{max}$ respectively. Additionally the size of the contained special interest groups is given.

Network	$\|V\|$	$\|E\|$	Avg.Deg.	APL	d	Density	C	#CC	$\|CC\|_{max}$	$\|KDML\|$	$\|WM\|$	$\|IR\|$	$\|ABIS\|$
LWA[≥0]∗	70	812	23.20	1.72	4	0.34	0.55	1	70	37	16	10	7
LWA[≥5]∗	65	227	6.99	2.53	5	0.11	0.33	1	65	34	15	9	7
LWA[≥10]∗	56	109	3.89	3.09	7	0.07	0.31	3	50	31	12	7	6

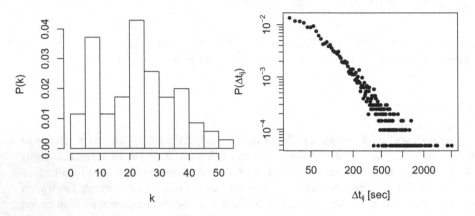

Fig. 1. Degree distribution $P(k)$ (left) and distribution of the different contact durations (right) in LWA[≥0]∗.

Furthermore, we extracted the "*visited talks*", i.e., the talks visited by each participant using the RFID information, resulting in 773 talk allocations for the conference participants.

4.3 Community Structure

LWA 2010 was a joint workshop week of four special interest groups of the German Computer Science Association (GI).

- **ABIS** focuses on *personalization and user modeling*.
- **IR** is concerned with *information retrieval*.
- **KDML** focuses on all aspects of *knowledge discovery and machine learning*.
- **WM**, for 'Wissensmanagement', considers all aspects of *knowledge management*.

During the registration for LWA 2010, each participant declared his affiliation to exactly one special interest group: KDML (37), WM (16), ABIS (7), IR (10), for a total of 70 participants. Since these interest groups capture common research interests as well as personal acquaintance, the set of participants is naturally clustered accordingly.

As a first characteristic for the interest groups, we aggregated the visited talks groupwise per track. Although several sessions were joint sessions of two interest groups, Figure 2 clearly shows for each group a strong bias towards talks of the associated conference track.

Table 2. Detail view on the distribution of the conference tracks of the talks visited by members of the different interest groups (numbers in percent). Members of the *ABIS* group, for example, visited ABIS talks in 58.8% of their visits, while they attended talks of the KDML track in 20.6%.

	ABIS	IR	KDML	WM
ABIS	**58.8**	13.2	20.6	7.4
IR	5.2	**60.4**	26.0	8.3
KDML	2.2	8.6	**80.2**	9.1
WM	16.3	14.1	26.1	**43.5**

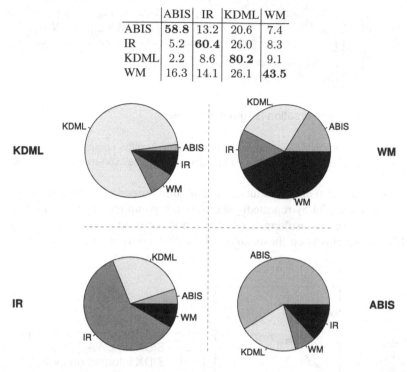

Fig. 2. Distribution of the conference tracks of the talks visited by members of the different interest groups. The top-left figure, for example, shows the distribution of tracks visited by the KDML special interest group.

Analysis of Community Dynamics. For our analysis of community dynamics, we focused on the contact graph using different minimal contact durations: In this respect, we analyzed, if an according community structure may be observed in the respective contact graphs obtained during the conference. We considered the modularity measure and the segregation index for assessing the respective community allocations.

Figure 3 shows the obtained weighted and unweighted modularity scores for the contact graphs $\mathsf{LWA}[\geq i]_\Sigma$ and $\mathsf{LWA}[\geq i]_\#$ with $i = 1, \ldots, 15$, considering the interest groups as communities. The results suggest consistent community interaction according to the modularity in the corresponding networks. For analyzing the impact of repeated or longer conversations, we calculated the weighted modularity score on the same networks, using either the number of conversations or the aggregated conversation time between two participants as edge weights. Figure 3 shows that the obtained modularity scores are nearly constant across the different networks. This is a first hint for assuming that peers tend to talk more frequently and longer within their associated interest groups. As we will see below, this hypothesis is confirmed by the segregation index analysis.

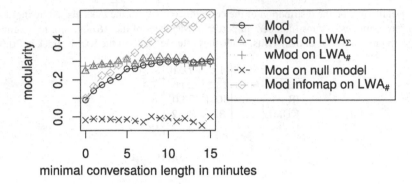

Fig. 3. Modularity scores for varying minimum conversation lengths, comparing the communities induced by the special interest groups to a null model, and to mined communities using Infomap.

To rule out statistical effects induced by structural properties of the contact graphs, we created a null model by repeatedly shuffling the group membership of all participants and averaging the resulting unweighted modularity scores. As Figure 3 shows, the shuffled group allocation shows no community structure in terms of modularity as expected.

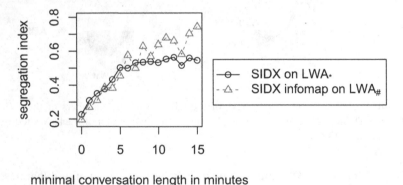

Fig. 4. Segregation index scores for varying minimum conversation lengths, comparing communities induced by the special interest groups to mined communities using Infomap.

Furthermore, we computed the segregation index for the communities on the unweighted graphs in order to exclude effects due to different network sizes. These could be introduced when filtering edges and nodes when increasing the minimum conversation length, cf., Table 1. As shown in Figure 4, the segregation index indicates a strongly ascending trend with increasing minimal conversation length concerning the corresponding networks and community allocations. This conforms to the intuition that more relevant (i. e., longer) conversations are biased towards dialog partners with common interests, as captured by the interest group membership.

Automatic Community Detection. Additionally, the standard community detection algorithm Infomap [25], which is shown to perform well [26], was chosen for reference and applied to the same contact graphs. Figure 3 also shows the unweighted modularity scores for the obtained (disjoint) communities. At a minimum conversation length of 5 minutes, it almost coincides with the modularity scores of the interest group community structure, and it nearly doubles both weighted and unweighted modularity scores in LWA$[\geq 15]_{\Sigma}$ and LWA$[\geq 15]_{\#}$.

Inspection of the obtained communities suggests that the applied algorithm yields communities in LWA$[\geq 0]_*$ which are similar to the given interest groups but also more specialized (i. e., sub communities) in LWA$[\geq i]_*$, $i \geq 5$. Figure 5 shows for reference in LWA$[\geq 0]_*$, that ABIS and IR are nearly perfectly captured by Infomap but KDML is split mainly across two communities, one of which shared with WM.

It is important to note in this context, that we do not aim at evaluating any community detection algorithm: We rather exemplify the application of such algorithms and approximate an upper bound of the modularity score in the contact graphs.

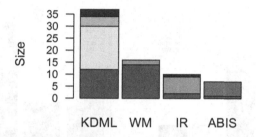

Fig. 5. Distribution of the interest groups across the six communities mined by the Infomap algorithm on LWA$[\geq 0]_*$. Each color corresponds to a single (non-overlapping) community.

Similar findings are obtained by applying the MOSES algorithm [27] for detecting overlapping communities. Figure 6 shows exemplary community detection results for minimum conversation lengths ($min = 0, 1, 5, 10$) using the MOSES algorithm.

The community results in Figure 6 suggest, that the communities tend to focus more on the special interest groups with an increasing minimum conversation length threshold: The communities start with a mixture of different interest groups (LWA$[\geq 0]_*$), but concentrate more and more on special sub-communities. These findings further confirm the initial results of Infomap and demonstrate the trend even more clearly, that more specialized sub-communities of the special interest groups are mined. This is especially significant for higher minimal conversation length thresholds, e. g., LWA$[\geq 10]_*$, see Figure 6. In summary, this community-oriented analysis suggests, that participants actually tend to interact more frequently with members of their own special interest group for longer conversations.

Community Densities. For analyzing social interactions across different interest groups we consider the density in the corresponding induced sub graphs.

Table 3 shows the density for each pair of interest groups $V_i, V_j \subseteq V$ in the complete contact graph $G = (V, E)$, the fraction of all actually realized edges in the set of possible edges between V_i and V_j.

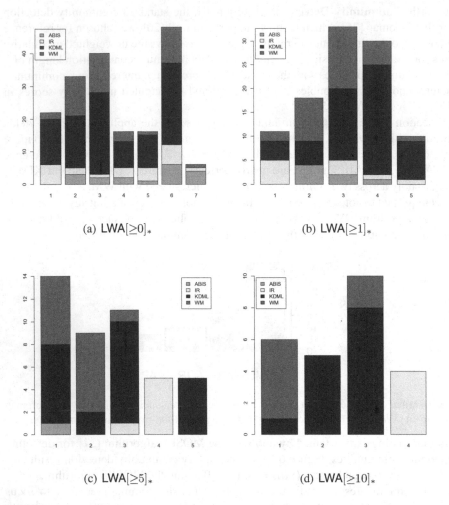

Fig. 6. Exemplary community detection results (overlapping communities) using the MOSES [27] algorithm. The different communities are colored according to their special interest track distributions.

Within the interest groups, the density values are strictly above the global density (cf., Table 1), but strictly below across different groups. This suggests that participants actually tend to interact more frequently with members of their own interest group.

Poster Session. As part of the conference, a joint poster session with accompanied buffet was scheduled. Typically, there is a lot of interaction at a poster session, leading to many contacts. Similar to the global inter-group densities in Table 3 the according inter-group densities for the poster session are shown in Table 4. Since the contacts of the poster session are a subset of all contacts for the whole conference, the density values are lower than those for the global densities.

Table 3. Density in the the contact graph LWA[≥0].

	ABIS	IR	KDML	WM
ABIS	**0.62**	0.23	0.19	0.28
IR	0.23	**0.44**	0.21	0.20
KDML	0.19	0.21	**0.38**	0.31
WM	0.28	0.20	0.31	**0.58**

However, we observe the same trends as for the global density, namely, that the intra-group densities are larger than the inter-group conversations – with one exception, i.e., the KDML special interest group. One possible explanation is, that KDML was the largest group present, and therefore the chances for increasing their inter-density shares with other special interest groups are better. Additionally, the large share of the density values compared to the global ones seem to indicate the impact and importance of the poster session for networking.

Table 4. Density in the complete contact graph of the poster session

	ABIS	IR	KDML	WM
ABIS	**0.19**	0.06	0.05	0.08
IR	0.06	**0.24**	0.07	0.08
KDML	0.05	0.07	0.10	**0.12**
WM	0.08	0.08	0.12	**0.36**

Discussion. In summary, our results exemplify, that community structure as induced by interest groups in conferences can be measured in face-to-face communication networks quite well. Furthermore, standard community detection algorithms show promising results for mining user communities from social interaction patterns in both on-line and face-to-face communication.

Typically, conference conversations tend to cluster around interest groups. For the LWA we could observe this effect for the present special interest groups. Intelligent analysis and mining techniques can then help, for example, for establishing more dynamic conference events.

4.4 Roles and Key Players

The assignment of roles to nodes in a network is a classification process that categorizes the players by common patterns. In this section, we discuss the connection between the academic status of the conference participants and the classic centrality measures and a community based role assignment. The latter was introduced in [28] together with the $rawComm$ measure that it is based on. The measure is defined as:

$$rawComm(u) = \sum_{v \in N(u)} \tau_u(v),$$

where the contribution of a node v in the neighborhood of u is given by:

$$\tau_u(v) = \frac{1}{1 + \sum_{v' \in N(u)} \left(I(v, v') * p + \bar{I}(v, v')(1 - q) \right)}.$$

Table 5. Group size and average graph centralities per academic position and for organizers and non-organizers in LWA[$\geq$5]$_*$: degree deg, strengths $str_\#$ and str_Σ, eigenvalue centralities eig_*, $eig_\#$ and eig_Σ, betweenness bet, closeness clo and the average community metric $rawComm$.

Position	size	deg	str$_\#$	str$_\Sigma$	eig$_*$	eig$_\#$	eig$_\Sigma$	bet	clo	rawComm
Prof.	10	7.500	16.700	11893.200	0.310	0.285	0.337	49.565	0.407	0.525
PostDoc	11	7.727	15.545	9793.364	0.303	0.213	0.198	75.973	0.419	0.675
PhD-student	33	7.152	15.091	9357.182	0.309	0.201	0.165	46.221	0.409	0.567
Student	5	3.600	12.400	6514.400	0.099	0.068	0.027	17.989	0.347	0.417
Other	6	6.667	14.333	8920.000	0.288	0.211	0.209	38.234	0.413	0.581
Organizer	11	10.000	23.727	15227.545	0.459	0.424	0.417	94.497	0.447	0.699
Non-Organizer	54	6.370	13.389	8408.056	0.256	0.162	0.144	39.565	0.397	0.542

In the formulas $N(u)$ is the neighborhood of a node u; $I(v, v') = 1$ if there is an edge between v and v' and 0 else; $\bar{I} = 1 - I$, p is the probability that an edge in the graph connects two communities and q is the probability that two non-linked nodes are in different communities. The $rawComm$ score can be interpreted as a measure of how much a node connects different communities (for details, cf., [28]). To estimate the probabilities p and q, we applied the method suggested in [28] using the contact graph and the special interest groups as grounding. The measure utilizes the (dynamically constructed) contact graphs since we focus on the time-dependent analysis of communities and roles, and the implied communication and interaction structure.

Although we considered a set of given communities as grounding (the special interest groups), it becomes clear by considering the results of the Infomap algorithm for automatic community detection (see Section 4.3) that some communities are subsets of the special interest groups, while others transcend the bounds of these groups. We therefore have chosen the probabilistic approach and used the clustering given through the special interest groups only for estimating the parameters p and q.

Global Characterization. Table 5 displays the average values of several graph node centralities of LWA[$\geq$ 5]$_*$ aggregated by academic position and for the conference organizers and non-organizers (regular conference participants), respectively. Note, that while the categories referring to academic status are disjoint (the category *other* includes all participants that do not fit one of the other four) organizers and non-organizers both include participants from all the 'status' categories.

A first observation is that the organizers have significantly higher scores in all nine measures under observation. In the considered conference scenario this is highly plausible due to the nature of an organizer's job during a conference – which in the case of LWA 2010 also included the supervision and maintenance of the RFID-experiment and the CONFERATOR. Among the four academic positions, striking differences can be noticed. First of all, the student scores in all centralities are lower than those of the other categories. We attribute this phenomenon to the fact, that students are less established in their scientific communities than scientists in higher academic positions and usually have little conference experience. This example motivates the use of social tools that could assist participants in initiating contact to their communities and persons of interest.

Within the categories "Prof.", "PostDoc" and "PhD-student" the eigenvalue centralities show a particular behavior. While the unweighted eigenvalue centrality eig_* does not fluctuate much, the weighted versions eig_Σ and $eig_\#$ increase strongly from one position to the next higher one. Eigenvalue centralities are considered a measure of importance. It seems plausible, that in a contact graph among scientists, the players with longer scientific experience – including a higher and broader degree of knowledge within scientific areas and more previous contacts and collaborations with their colleagues – are considered more important and that this attitude is reflected in their contacts. The node strength measures show similar results. While the degree deg is only slightly different among the three positions, the weighted versions $str_\#$ and especially str_Σ show large differences and increase together with the position. The considerable difference between the weighted and unweighted measures indicates the relevance of the frequency and the length of the contacts: Professors, for example, have longer and more contacts to other participants than PostDocs.

Another aspect is illustrated by the betweenness (bet) scores: Relatively to the other groups, a lot of shortest paths of LWA$[\geq 5]_*$ run through nodes of PostDocs. We attribute this to the structure of scientific institutes, where usually one professor supervises several PostDocs who again each supervise several PhD-students. PostDocs are the connection between professors and the postgraduates and thus assume the role of gatekeepers in their working environment.

Explicit Roles. For the $rawComm$ measure it is harder to come up with a plausible explanation for the different values and the order of the academic positions. However, as described in [28], it can be combined with $ndeg$ – the degree divided by the maximum degree – to gain a role classification for the network's nodes in the following way: One out of four roles is assigned to a node v according to

$$
role(v) := \begin{cases}
\text{Ambassador} & ndeg(v) \geq s,\, rawComm(v) \geq t \\
\text{Big Fish} & ndeg(v) \geq s,\, rawComm(v) \lesssim t \\
\text{Bridge} & ndeg(v) \lesssim s,\, rawComm(v) \geq t \\
\text{Loner} & ndeg(v) \lesssim s,\, rawComm(v) \lesssim t
\end{cases}
$$

where s and t are thresholds that we chose as $s = t = 0.5$ – the same choice as in [28].

Ambassadors are characterized by high scores in both degree and $rawComm$ which means that they connect many communities in the graph. A *Big Fish* has contacts to a lot of other nodes, however, mostly within the same community. *Bridges* connect communities, however, not as many as ambassadors. Finally, *Loners* are those with low scores in both measures.

In the following, we investigate how nodes in their explicitly given roles like the academic position and the job (organizer) fill those implicitly given graph structure-based roles. Therefore, we applied the role classifier to the graphs LWA$[\geq 0]_*$ through LWA$[\geq 15]_*$ to determine – under the assumption, that longer contacts indicate more serious and scientific discussions – how this changes the community roles.

The first immediate finding is, that in none of the graphs any participant was ever classified as Big Fish, i. e., whenever a node has a high degree it also has a high

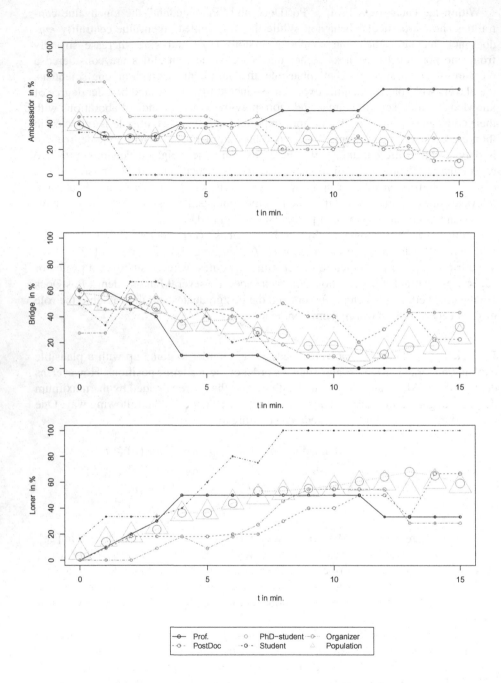

Fig. 7. Fraction of participants that assume the roles Ambassador, Bridge and Loner

rawComm score. We attribute this peculiarity to the fact, that the very nature of social interaction at conferences usually is exchanging ideas with participants outside the own peer group. Especially during the LWA 2010, participants were encouraged to engage in interdisciplinary dialogue for example by including several joint sessions in the schedule and a combined event of social dinner and poster session.

The first of the three diagrams in Figure 7 displays the percentage of participants with a common academic position or job that were classified as Ambassador. The line marked with triangles displays that fraction of all participants together. The second and third diagram display the same fractions for the roles Bridge and Loner. For example in LWA[≥ 0]$_*$, 40% of the professors were classified as Ambassador, 60% as Bridge and 0% as Loner.

In each diagram the size of the nodes indicates the size of the group of participants with the examined position/job in the respective graph. The PhD-students, for example, are the largest section, while the students form the smallest. For LWA[≥ 5]$_*$ those sizes are given in Table 5. While all curves in Figure 7 fluctuate, there are several clearly visible tendencies. In all three diagrams, the fractions of PhD-students is very close to the fraction of all participants. The simple reason for that is, that PhD-students are the majority within the conference population and therefore dominate the general behavior.

Many of the organizers start out as Ambassador or Bridge. This is again consistent with their job description. However, filtering out short contacts and thus the typical quick organizational conversations, the relevance of the organizers decreases with a higher limit to the minimum contact length. More and more organizers become Loners; in the last graph LWA[≥ 15]$_*$, they are almost equally distributed among the three roles. One should keep in mind, that the group of organizers contains persons in all academic positions. Therefore, after filtering out most of the contacts that presumably are due to their organizational work, the organizers act mainly in their respective role as conference participants, which might explain the stronger fluctuations in the right part of the curve.

Very consistent with the findings described above is the role distribution among the students. While in the first graphs, where short contacts dominate the longer ones, some of them are classified as Bridge or Ambassador, they quickly disappear from those roles and are classified as Loner. Compared to the PhD-students, the fractions of the PostDocs are with few exceptions higher for the roles Ambassador and Bridge and lower for Loner. This is again consistent with the previous observations concerning the graph structure measures. Due to their greater experience PostDocs seem to have more access to colleagues in other communities. However, with the increasing filter limit, like most of the participants they become classified as loners. Finally, the curve of the professors in the role Ambassador shows the largest deviation from the mainstream. While in that role all other group's fractions decrease, that of the professors increases significantly up to 70% which is far more than any of the other academic positions.

In summary, we observe, that the chosen method of role assignment seems to be highly correlated to the roles like academic position and the organizer job.

Characterization of Explicit Roles. In the following, we aim to characterize the roles in more detail; for the dataset, we focus on the majority classes, i.e., we consider the target concept *non-organizer* concerning roles, and the target concept *PhD-students*

Table 6. Role = Non-Organizer / Position = PhD-student for the aggregated count information with an aggregated contact length $\geq$ 5 min. The tables show the lift of the pattern comparing the fraction of non-organizers / PhD-students covered by the pattern p compared to the fraction of the whole dataset, the size of the pattern extension (number of described non-organizers / PhD-students), and the description itself.

target	#	lift	p	size	description
	1	1.06	0.88	51	$clo=\{low; medium\}$
	2	1.05	0.87	61	$eig_*=\{low; medium\}$
Non-Organizer	3	1.04	0.86	59	$deg=\{low; medium\}$
	4	1.10	0.92	12	$clo=\{low; medium\}$ AND $deg=\{high; medium\}$
	5	1.12	0.93	30	$clo=\{high; low\}$ AND $eig_*=\{low; medium\}$
	1	1.07	0.54	59	$bet=\{high; low\}$
	2	1.07	0.54	48	$s=\{high; low\}$
PhD-student	3	1.14	0.58	26	$deg=high$
	4	1.31	0.67	12	$bet=\{high; low\}$ AND $eig_*=high$
	5	1.38	0.70	20	$deg=high$ AND $bet=\{high; low\}$
	6	1.58	0.80	10	$deg=high$ AND $bet=\{high; low\}$ AND $eig_*=\{high; low\}$

concerning academic position. For the analysis, we applied a method for mining characteristic patterns [29] based on subgroup discovery techniques, e.g., [30,31,32]: Basically, we aim to discover subgroups of participants described by combinations of factors, e.g., *deg=high AND bet=high* that show a high share of a certain target property, e.g., with respect to organizing or academic roles. For the data preprocessing, we first discretized the numeric features described above into three intervals (low, medium, high) using equal-width discretization.

The most descriptive factors for the role *non-organizer* are shown in Table 6 (upper). They confirm the averaged results shown above, in that the most characteristic *single* factors are given by the closeness, eigenvalue centrality, and the degree of the non-organizers, for which lower values than those of the organizers are measured. However, if we consider combinations of factors, we observe, that there are subgroups regarding the role non-organizer for which extreme values, e.g., of the closeness together with the eigenvalue centrality yield a significant increase in characterization power, as shown by the quality increase in Table 6.

If we consider the largest group *PhD-student* (concerning the academic positions), we observe the single factors shown in Table 6 (lower), also confirming the averaged results presented above. Similarly to the non-organizers, we see that extreme values, i.e., sets of high and low values, are also very significant for distinguishing PhD students. As expected the combination with other strong influence factors increases the precision of the patterns (indicated by the *lift* parameter).

5   Conclusions

In this paper, we have presented results of an in-depth analysis of the dynamics of community structure and roles of face-to-face contacts during a conference. We have performed various analyses on data collected during the LWA 2010 in Kassel in October

2010 by using a social conference guiding system. We analyzed and described high-level statistics of the collected network data, examined the different communities, the roles and key players concerning these and the conference in total. The analysis was grounded using real-world conference data capturing community information about participants (membership in special interest groups). It was shown that there is detectable community structure in the investigated face-to-face networks that is consistent with the one given through the groups. Further, the structural properties of the contact graphs obtained at the LWA conference reflect different aspects of community interactions and roles. For the latter, we provided characteristic (role) profiles and detected traces of different behavioral patterns for different roles.

As future work, we plan to continue the community related investigations further, since communities play a central role for a social conferencing system and should allow and support emergence and evolution of community structure. Furthermore, identifying key actors according to their roles is an interesting task; results could be used e.g., for creating virtual sessions or recommendations. Additionally, we aim to collect more data in order to generalize and compare the mined profiles and patterns on a broader basis, e.g., concerning more diverse conferences. A first step in this direction has been taken in [33], continuing these analyses seems a very promising task.

Acknowledgements. This work has been supported by the VENUS research cluster at the interdisciplinary Research Center for Information System Design (ITeG) at Kassel University. CONFERATOR applies active RFID technology which was developed within the SocioPatterns project, whose generous support we kindly acknowledge. We also wish to thank Milosch Meriac from Bitmanufaktur in Berlin for helpful discussions regarding the RFID localization algorithm. Our particular thanks go the SocioPatterns team, especially to Ciro Cattuto, who enabled access to the Sociopatterns technology, and who supported us with valuable information concerning the setup of the RFID technology.

References

1. Wongchokprasitti, C., Brusilovsky, P., Para, D.: Conference Navigator 2.0: Community-Based Recommendation for Academic Conferences. In: Proc. Workshop Social Recommender Systems, IUI 2010 (2010)
2. Atzmueller, M., Benz, D., Doerfel, S., Hotho, A., Jäschke, R., Macek, B.E., Mitzlaff, F., Scholz, C., Stumme, G.: Enhancing Social Interactions at Conferences. IT - Information Technology 53(3), 101–107 (2011)
3. Alani, H., Szomszor, M., Cattuto, C., Van den Broeck, W., Correndo, G., Barrat, A.: Live Social Semantics. In: Bernstein, A., Karger, D.R., Heath, T., Feigenbaum, L., Maynard, D., Motta, E., Thirunarayan, K. (eds.) ISWC 2009. LNCS, vol. 5823, pp. 698–714. Springer, Heidelberg (2009)
4. Hui, P., Chaintreau, A., Scott, J., Gass, R., Crowcroft, J., Diot, C.: Pocket Switched Networks and Human Mobility in Conference Environments. In: Proc. 2005 ACM SIGCOMM Workshop on Delay-Tolerant Networking, WDTN 2005, pp. 244–251. ACM, New York (2005)
5. Eagle, N., Pentland, A.S.: Reality Mining: Sensing Complex Social Systems. Personal Ubiquitous Comput. 10, 255–268 (2006)

6. Meriac, M., Fiedler, A., Hohendorf, A., Reinhardt, J., Starostik, M., Mohnke, J.: Localization Techniques for a Mobile Museum Information System. In: Proceedings of WCI (Wireless Communication and Information) (2007)

7. Cattuto, C., den Broeck, W.V., Barrat, A., Colizza, V., Pinton, J.F., Vespignani, A.: Dynamics of Person-to-Person Interactions from Distributed RFID Sensor Networks. PLoS ONE 5(7) (July 2010)

8. Isella, L., Stehlé, J., Barrat, A., Cattuto, C., Pinton, J.F., den Broeck, W.V.: What's in a crowd? analysis of face-to-face behavioral networks. Journal of Theoretical Biology 271(1), 166–180 (2011)

9. Barrat, A., Cattuto, C., Szomszor, M., Van den Broeck, W., Alani, H.: Social Dynamics in Conferences: Analyses of Data from the Live Social Semantics Application. In: Patel-Schneider, P.F., Pan, Y., Hitzler, P., Mika, P., Zhang, L., Pan, J.Z., Horrocks, I., Glimm, B. (eds.) ISWC 2010, Part II. LNCS, vol. 6497, pp. 17–33. Springer, Heidelberg (2010)

10. Stehle, J., Voirin, N., Barrat, A., Cattuto, C., Isella, L., Pinton, J.F., Quaggiotto, M., den Broeck, W.V., Regis, C., Lina, B., Vanhems, P.: High-resolution measurements of face-to-face contact patterns in a primary school. CoRR abs/1109.1015 (2011)

11. Isella, L., Romano, M., Barrat, A., Cattuto, C., Colizza, V., den Broeck, W.V., Gesualdo, F., Pandolfi, E., Rava, L., Rizzo, C., Tozzi, A.E.: Close encounters in a pediatric ward: measuring face-to-face proximity and mixing patterns with wearable sensors. CoRR abs/1104.2515 (2011)

12. Brandes, U., Erlebach, T. (eds.): Network Analysis. LNCS, vol. 3418. Springer, Heidelberg (2005)

13. Chou, B.-H., Suzuki, E.: Discovering Community-Oriented Roles of Nodes in a Social Network. In: Bach Pedersen, T., Mohania, M.K., Tjoa, A.M. (eds.) DAWAK 2010. LNCS, vol. 6263, pp. 52–64. Springer, Heidelberg (2010)

14. Lerner, J.: Role Assignments. In: Brandes, U., Erlebach, T. (eds.) Network Analysis. LNCS, vol. 3418, pp. 216–252. Springer, Heidelberg (2005)

15. Scripps, J., Tan, P.-N., Esfahanian, A.-H.: Node Roles and Community Structure in Networks. In: Proc. 9th WebKDD and 1st SNA-KDD 2007 Workshop on Web Mining and Social Network Analysis, pp. 26–35. ACM, New York (2007)

16. Diestel, R.: Graph theory. Springer, Berlin (2006)

17. Gaertler, M.: Clustering. In: [12], pp. 178–215

18. Newman, M.E., Girvan, M.: Finding and Evaluating Community Structure in Networks. Phys. Rev. E Stat. Nonlin. Soft. Matter Phys. 69(2), 026113.1–026113.15 (2004)

19. Mitzlaff, F., Benz, D., Stumme, G., Hotho, A.: Visit Me, Click Me, Be My Friend: An Analysis of Evidence Networks of User Relationships in Bibsonomy. In: Proceedings of the 21st ACM Conference on Hypertext and Hypermedia, Toronto, Canada (2010)

20. Leskovec, J., Lang, K.J., Mahoney, M.W.: Empirical Comparison of Algorithms for Network Community Detection, cite arxiv:1004.3539 (2010)

21. Chin, A., Chignell, M.: Identifying Communities in Blogs: Roles for Social Network Analysis and Survey Instruments. Int. J. Web Based Communities 3, 345–363 (2007)

22. Freeman, L.: Segregation In Social Networks. Sociological Methods & Research 6(4), 411 (1978)

23. Newman, M.E.J.: Analysis of Weighted Networks (2004), http://arxiv.org/abs/cond-mat/0407503

24. Nicosia, V., Mangioni, G., Carchiolo, V., Malgeri, M.: Extending the Definition of Modularity to Directed Graphs with Overlapping Communities. J. Stat. Mech. (2009)

25. Rosvall, M., Bergstrom, C.: An Information-theoretic Framework for Resolving Community Structure in Complex Networks. Proc. Natl. Acad. of Sciences 104(18), 7327 (2007)

26. Lancichinetti, A., Fortunato, S.: Community Detection Algorithms: A Comparative Analysis, arxiv:0908.1062 (2009)
27. McDaid, A., Hurley, N.: Detecting highly overlapping communities with model-based overlapping seed expansion. In: Proceedings of the 2010 International Conference on Advances in Social Networks Analysis and Mining, ASONAM 2010, pp. 112–119. IEEE Computer Society, Washington, DC (2010)
28. Scripps, J., Tan, P.N., Esfahanian, A.H.: Exploration of Link Structure and Community-Based Node Roles in Network Analysis. In: ICDM, pp. 649–654 (2007)
29. Atzmueller, M., Lemmerich, F., Krause, B., Hotho, A.: Who are the Spammers? Understandable Local Patterns for Concept Description. In: Proc. 7th Conference on Computer Methods and Systems (2009)
30. Wrobel, S.: An Algorithm for Multi-Relational Discovery of Subgroups. In: Komorowski, J., Żytkow, J.M. (eds.) PKDD 1997. LNCS, vol. 1263, pp. 78–87. Springer, Heidelberg (1997)
31. Atzmueller, M., Puppe, F., Buscher, H.-P.: Exploiting Background Knowledge for Knowledge-Intensive Subgroup Discovery. In: Proc. 19th Intl. Joint Conference on Artificial Intelligence (IJCAI 2005), Edinburgh, Scotland, pp. 647–652 (2005)
32. Atzmüller, M., Puppe, F.: SD-Map – A Fast Algorithm for Exhaustive Subgroup Discovery. In: Fürnkranz, J., Scheffer, T., Spiliopoulou, M. (eds.) PKDD 2006. LNCS (LNAI), vol. 4213, pp. 6–17. Springer, Heidelberg (2006)
33. Macek, B.-E., Scholz, C., Atzmueller, M., Stumme, G.: Anatomy of a Conference. In: Proc. 23rd ACM Conference on Hypertext and Social Media. ACM Press (2012)

Factors Influencing the Co-evolution of Social and Content Networks in Online Social Media

Philipp Singer[1], Claudia Wagner[2], and Markus Strohmaier[3]

[1] Knowledge Management Institute, Graz University of Technology, Graz, Austria
philipp.singer@tugraz.at
[2] DIGITAL Intelligent Information Systems,
JOANNEUM RESEARCH, Graz, Austria
claudia.wagner@joanneum.at
[3] Knowledge Management Institute and Know-Center,
Graz University of Technology, Graz, Austria
markus.strohmaier@tugraz.at

Abstract. Social media has become an integral part of today's web and allows communities to share content and socialize. Understanding the factors that influence how communities evolve over time - for example how their social network and their content co-evolve - is an issue of both theoretical and practical relevance. This paper sets out to study the temporal co-evolution of microblog messages' content and social networks on Twitter and of forum-messages' content and social networks induced from communication behavior of users from an online forum called Boards.ie and bi-directional influences between them by using multilevel time series regression models. Our findings suggest that social networks have a stronger influence on content networks in our datasets over time than vice versa, and that social network properties, such as Twitters users' in-degree or Boards.ie users' reply behavior, strongly influence how active and informative users are. While our investigations are limited to three small datasets obtained from Twitter and Boards.ie, our analysis opens up a path towards more systematic studies of network co-evolution on social media platforms. Our results are relevant for researchers and community managers interested in understanding how content-related and social behavior of social media users evolve over time and which factors impact their co-evolution.

Keywords: Microblog, Twitter, Boards.ie, Influence Patterns, Semantic Analysis, Time Series.

1 Introduction

Social media applications such as blogs, message boards or microblogs allow communities of users to share content and socialize. Managing and hosting such communities can however be a costly and time consuming task, and community managers and hosts need to ensure that their communities remain active and popular. Monitoring and analyzing behavior of online communities and their

M. Atzmueller et al. (Eds.): MSM/MUSE 2011, LNAI 7472, pp. 40–59, 2012.

co-evolution over time can provide valuable information on the factors that impact the activity and popularity of such communities. Activity and popularity of a community are often measured by the growth of content produced by the community and/or the growth of its social network. However, as a research community we know little about the factors that impact the activity of a community on social media applications and we know even less about how users' social behavior (i.e., their following or chatting behavior) influence their content-related activities (e.g., their authoring or tagging behavior) and vice versa.

This paper sets out to explore factors that impact the co-evolution of communities' content-related and social behavior. Specifically, we explore the temporal co-evolution of content and social networks in Twitter, a popular microblogging platform, and Boards.ie[1], a popular Irish message board. This work represents an extended version of a paper titled *Understanding co-evolution of social and content networks on Twitter* published at the workshop *MSM2012: Making Sense of Microposts* of the *WWW2012 conference* [17]. It adds new information by presenting results from two additional experiments conducted on data from Twitter and Boards.ie. In this work we use two different Twitter datasets which both contain information about users' daily content-related activities and outcomes as well as information about their explicitly defined social network and related activities. The first Twitter dataset is based on randomly chosen users taken from Twitter's public timeline and the second Twitter dataset is formed by community seed users taken from an user list. Both datasets were crawled within a temporal window of 30 days. Further we use one dataset obtained from Boards.ie which contains all messages published between July 2005 and December 2006. From those messages we constructed a social network of authors by analyzing their reply-behavior of users (i.e., we constructed a communication-partner networks where user A is related with user B if A previously replied to B).

By using multilevel time series regression models, we aim to identify factors that explain how Twitter's communication content networks and social networks co-evolve over time and to find influences between reply-behavioral properties and the content related topical focus of Boards.ie users. Time series modeling is a way to forecast and predict future values based on previously measured values. Such models can help to reveal influences between variables over time. There are several possible ways to model time series appropriately and we chose an *autoregressive multilevel regression model* in order to identify time based influences in our datasets. We decided to use such a model instead of a non-autoregressive one because this allows us to explore changes over time.

Unlike previous research, we focus on measuring dynamic bi-directional influence between these networks in order to identify which content-related factors impact the evolution of social networks and vice versa. This analysis enables us to tackle questions such as:

- *Does growth of a Twitter user's followers increase the number of links or hashtags they use per tweet?*

---

[1] http://www.boards.ie

- *Does an increase in Twitter users' popularity imply that their tweets will be retweeted more often on average?*
- *Do Twitter users who share more links or use more hashtags gain more followers?*
- *Do the number of own replies to other posts or the number of replies by other users influence the topical focus of Boards.ie users?*

Our results reveal interesting insights into influence patterns in content networks, social networks and between them. For example, our results highlight the importance of social networks on our Twitter datasets and show that social network properties such as users' *in-degree* strongly influence how active users are (i.e., their *number of tweets*) in our sample of Twitter users. That means if the number of followers of a user increases, also the number of tweets he/she publishes increases the next day. Beside that the Twitter users' in-degree also has a positive influence on their *retweeted ratio* and *link ratio*, which means that an increase of users' followers implies that users become more informative and that their tweets are retweeted more often by others. Furthermore, our experiments on the Boards.ie dataset point out that active users in this dataset tend to become more active over time and that the users extend their repertoire of topics used throughout their posts depending on how much their number of communication partners increased previously.

Our observations and implications are relevant for researchers interested in social network analysis, text mining and behavioral user studies, as well as for community manager and hosts which need to understand the factors that influence the evolution of their communities in terms of their content-related and social behavior.

This paper is organized as follows: Chapter 2 discusses related research on exploring influence patterns in social media. In chapter 3 we describe our methodology. Chapter 4 presents the three datasets to which we applied our method. Our experimental setup is described in chapter 5 and finally our results are discussed in chapter 6. We conclude our work, discuss limitation of our work and give an outlook on our planned future work in chapter 7.

2 Related Work

The study of influence has a long history in the fields of sociology, communication, marketing and political science [5]. Influence detection in social media has attracted a lot of attention within the last few years and focused mainly on exploring factors that influence users' behavior and finding influential users.

The problem of detecting influential users for certain topics has for example been studied by Cha et al. [5] in the context of Twitter. They defined an influential user as someone who has the potential to lead others to engage in a certain act. Their results highlight that across all three processed measures (following other users, sharing tweets, responding to other tweets), the top influentials were mostly well-known public figures or websites, but there was nearly no overlap between the three top lists. Kwak et al. [13] identified influential users by rank-ing them based on three different measures: number of followers, page rank and number of retweets. The authors showed that the ranking of influential Twit-ter users differs on the approached measure. Especially the ranking on retweets differs from the other two rankings. Bakshy et al. [3] investigated the attributes and relative influence of Twitter users by tracking diffusion events. The results show that the largest cascades are produced by users, who have already been in-fluential in the past and also have a large number of followers. Furthermore, the observations point out that the most influential users also tend to be the most cost-effective ones. Our work differs from previously described research since we do not try to identify influential users, but rather find factors that influence users' behavior in various ways.

The problem of identifying factors which influence users' behavior has amongst others been studied by Rowe and colleagues [16] on Youtube data. They analyzed correlations between several content features and user-behavior based features. Their results show that only the post view count has a significant correlation with the subscriber count of a user. Their work further demonstrates that it is possi-ble to accurately predict the subscriber count by detecting an optimal combina-tion of features. Anagnostopoulos et al. [2] explored the influence of Flickr's social network on the content based tagging behavior of users. They used two different statistical techniques (shuffle test and edge-reversal test) to reveal potential influ-ences. Their results show that few social influences exist on Flickr, which indicates that social networks in Flickr are less important than in Twitter.

The approach of finding such influence patterns in social media has also been studied in Twitter. Suh et al. [19] studied factors which influence the retweet-ability of tweets. Unlike above described papers, the authors tried to find social as well as content factors that affect a content property. The results show that URLs and hashtags are significant factors impacting the retweetability, whereas the domain of the URL and the type of hashtag matter. Unlike the work by Cha et al. [5] their results point out that the number of followers (i.e., the in-degree of a user) has a positive effect on the retweetability of his/her tweets. Interest-ingly, the number of tweets an author published in the past does not have an important impact on his/her retweetability. Similar work was done by Naveed et al. [15] who explored the relation between content properties of tweets and the likelihood of tweets being retweeted. By analyzing a logistic regression model's coefficients, the authors found that the inclusion of a hyperlink and using terms of a negative valence increases the likelihood of the tweet being retweeted. We have shown in previous work that motivational and behavioral factors also influ-ence properties of social media, such as emergent semantics, the degree to which users agree on tags [12] or the suitability for classification [25].

Unlike previous research we do not focus on identifying the most popular or influential content or user by analyzing users' content and social network, but aim to understand how these networks evolve, which factors influence their evolution and which bi-directional influences exist. Nevertheless, we are not the first to study the longitudinal influence between social and content properties (see for example [1] or [6]. However, our work differs from previous work since we study the longitudinal influence between social and content network properties explicitly in a social network environment based on Twitter and a public online message board called Boards.ie. In comparison to most of the recent work we also do not only focus on one direction of influence (e.g., if social aspects influence content properties), rather we provide insights into bi-directional influences as well as a self-influence detection.

The idea of this paper is based on research conducted by Wang and Groth [22] who present a multilevel autoregressive regression model to detect the bi-directional influences between social and content parameters. The authors were one of the first to analyze both, communication behavior of users together with the related outcome of these communications, the content.

Our work differs from their work in the sense that we additionally are exploring influences between communication content networks and explicit social networks on Twitter, while Wang and Groth investigated relations between publications' content and co-authorship networks and between communication networks and communication content in forums.

3 Methodology

Since we aim to gain insights into the temporal evolution of content networks and social networks, we apply *time series modeling* [11] based on the work by Wang and Groth [22] who provided a framework to measure the bi-directional influence between social and content network properties. We applied an *autoregressive model* in order to model our time series data. An autoregressive model is a model that goes back p time units in the regression and has the ability to make predictions. This model can be defined as $AR(p)$, where the parameter p determines the order of the model. An autoregressive model aims to estimate an observation as a weighted sum of previous observations, which is the number of the parameter p. In this work we apply a simple model, which calculates each variable independently and further only includes values from the last time unit. We also focus as a first step on a linear autoregressive model. It can be very well the case that such nonlinear models may be more suitable for our analyses, because such influence dynamics in our datasets may have nonlinear behavior. We leave this investigation of other nonlinear models open for future work. The calculated statistical-significant coefficients of the model can determine the influences between variables over time.

In regression analysis variables often stem from different levels. So called *multilevel regression models* are an appropriate way to model such data. Hence, the measurement occasion is the basic unit which is nested under an individual, the cluster unit. In our datasets we have such a hierarchical nested structure. Each day different properties are measured repeatedly, but all of these values belong to different individuals in our study [10]. If we would apply a simple autoregressive model to our data we would ignore the difference between each user and would just calculate the so-called *fixed effects*, because we can not assume that

all cluster-specific influences are included as covariates in the analysis [18]. The
advantage of such multilevel regression models is now that they add *random effects* to the fixed effects to also consider variations among our individuals. Since
we measure different properties repeatedly for different days and different individuals in our study, our dataset has a hierarchical nested structure. Therefore,
we utilize a *multilevel autoregressive regression model* which is defined as follows:

$$x_{i,p}^{(t)} = a_i^T x_p^{(t-1)} + \epsilon_i^{(t)} + b_{i,p}^T x_p^{(t-1)} + \epsilon_{i,p}^{(t)} \tag{1}$$

In this equation $x_p^{(t)} = (x_{i,p}^{(t)}, ..., x_{m,p}^{(t)})^T$ represents a vector, which contains the
variables for an individual p at time t. Furthermore, $a_i = (a_{i,1}, ..., a_{im})^T$ represents the fixed effect coefficients and $b_i = (b_{i,1}, ..., b_{im})^T$ represents the random
effect coefficients. It is assumed that $\epsilon_i^{(t)}$ and $\epsilon_{i,p}^{(t)}$ is the noise with Gaussian
distribution for the fixed and random effects respectively. It has zero mean and
variance σ_ϵ^2. To compare the fixed effects to each other, the variables in the
random effect regression equations need to be linearly transformed to represent
standardized values (see section 5.2). How this model is finally applied to our
data is described in section 5.4.

4   Datasets

We explore three different datasets in this work, a randomly-sampled Twitter
dataset, a community-centered Twitter dataset and an online forum dataset obtained from Boards.ie. We chose these three representative datasets because we
want to explore phenomenons in different application setting and want to contrast our experiments for different groups of seed users in the same social media
application. The Twitter datasets were both crawled within a time period of 30
days including 30 time steps for each day, whereas our Boards.ie dataset covers
twelve time steps for each beginning of a month in 2006. Twitter is a popular
micro-blogging service and allows communities of users to share content, communicate, create explicit social networks and form groups. Boards.ie is a very
well-known Irish message forum which gives users the possibility to communicate within 725 forums about a diverse range of topics such as movies or specific
video games. While Twitter allows studying the co-evolution of microblog messages' content and explicit social networks, Boards.ie enables us to study the
co-evolution of forum-messages' content and social networks which were induced
from communication behavior of users (i.e., a communication network of users
where user A is related with user B if A replied to a message authored by B).

Table 1 shows the different property values of all three datasets averaged over
all users and time steps (30 for Twitter and 12 for Boards.ie). The daily/monthly
measurements cover social (i.e., followers and followees of users for Twitter and
reply behavior for Boards.ie) properties as well as content (i.e., posts of users)
properties. The three datasets differ in their domain, crawling strategies, requirements, network shapes and measured time steps and are described in the
following sections.

Table 1. Properties of each dataset averaged over all users (134 users for the community dataset, 1,188 users from the randomly-sampled dataset and 29,886 users from the Boards.ie dataset) and all time steps; ID = in-degree, OD = out-degree, ID-C = in-degree centrality, OD-C = out-degree centrality, BC = betweenness centrality, CC = clustering coefficient, HR = Hashtag ratio, RTR = Retweet ratio, RTDR = Retweeted Ratio, LR = Link ratio, NT = Number of tweets, TE = Topic Entropy

	Social network properties						Content network properties					
Dataset	OD	ID	OD-C	ID-C	BC	CC	HR	RTR	RTDR	LR	NT	TE
Randomly-Sampled	691.65	1673.06	x	x	x	x	0.179	0.065	0.1	0.152	44.622	x
Community	x	x	0.198	0.198	116.771	0.497	0.1974	0.081	0.067	0.202	2.610	x
Boards.ie	11.38	11.34	x	x	x	x	x	x	x	x	x	0.355

4.1 Randomly-Sampled Dataset

To generate the first dataset, we randomly chose 1,500 users from the public Twitter timeline who we used as *seed users*. We used the public timeline method from the Twitter API to sample users rather than using random user IDs since the timeline method is biased towards active Twitter users. To ensure that our random sample of seed users consists of active, english-speaking Twitter users, we further only kept users who mainly tweet in English, have at least 80 followers, 40 followees and 200 tweets. We also had to remove users from our dataset who deleted or protected their account during the 30 days of crawling. Hence, we ended up having 1,188 seed users for whom we were able to crawl their social network (i.e., their followers and followees) and their tweets and retweets published on each day within a 30 day time period (from 15.03.2011 to 14.04.2011). The constructed social network of seed users only reflects a sub-part of a greater network. Therefore it makes no sense to calculate and analyze specific network properties such as betweenness centrality or clustering coefficient, because these properties depend on the whole network and we only have data available for a certain sub-network.

4.2 Community Dataset

To generate the second dataset, we decided to use a community of users as *seed users* rather than using random users. We used a Twitter user list[2] called *semweb* which was set up by Stefano Bertolo[3], who is a professional in knowledge representation, information retrieval, natural language representation and

---

[2] http://twitter.com/#!/sclopit/semweb
[3] http://twitter.com/#!/sclopit

related topics. Stefano Bertolo is a researcher working for the European Commission who is amongst others responsible for the research area *semantic web*. We assume that the users in his list are relevant for the topic semantic web and belong to the semantic web community.

We used all publicly available users from this list (134 users out of 137 in total) and used them as seed users. For all 134 seed users, we crawled their social network (i.e., their followers and followees) and all their tweets and retweets published on each day within the 30 day time period (from 26.04.2011 to 25.05.2011). Since our seed users of this dataset belong to a community, it is likely that they are heavily connected via follow-relations on Twitter. Note that we collected edges between our seed users and discarded edges pointing or coming from Twitter users outside this group of users. That means we created a sub-network of users belonging to a community, which might therefore be linked together with lots of edges.

4.3   Boards.ie Dataset

This third dataset is based on a collected subset of a popular Irish message board called Boards.ie[4] for a total of 29,886 distinct users. We focus on properties calculated for the year 2006.

At first, all posts in the time period between the beginning of July 2005 and the end of 2006 were extracted. For each user and each month in 2006 we aggregated all posts which were authored by this user within the last 6 months. Based on this monthly aggregates of user messages we created users' content network and social network for the current month. Unlike Twitter, Boards.ie does not provide explicit social relations between user. Therefore, we built the Boards.ie social network for users by weighting edges cumulatively by the number of replies between any two users. Since the content of each user in this Boards.ie dataset is substantially longer than in our Twitter dataset, we use topic models to abstract users' content aggregation and represent them as a distribution of topics rather than a collection of words.

In order to learn topics from a collection of documents and annotate users with topics we fitted a *Latent Dirichlet allocation* (LDA) model [4] to our dataset. LDA is a generative model for document collections and it models the generation of a document collection as a three step process: First, for each document d a distribution over topics θ_d is sampled from a *Dirichlet distribution* α. Second, for each word w_d in the document d, a single topic z is chosen according to this distribution θ_d. Finally, each word w_d is sampled from a multinomial distribution over words ϕ_z which is specific for the sampled topic z.

Fitting a LDA model to a collection of trainings documents requires finding the parameters which maximize the posterior $P(\theta, \phi, z | \alpha, \beta, w, .)$. In our experiments we used MALLET's [14] LDA implementation, selected the standard hyperparameter of LDA ($\alpha = 50/T$ and $\beta = 0.01$) and alternatively optimized the topic parameters given the hyperparameters, and the hyperparameters given

---

[4] http://http://www.boards.ie/

the topic parameters. Based on the empirical findings of [21], we decided to place an asymmetric Dirichlet prior over the topic distributions and a symmetric prior over the distribution of words. According to our initial intuition we set the number of topics $T = 80$. Following the model selection approach described in [9] we selected the number of topics by evaluating the probability of held-out data for various values of T. On average (over 10 runs) the model gives the highest probability to heldout documents if $T = 650$.

That means the content network of the Boards.ie dataset consists of user nodes and 650 topic nodes which are connected via weighted edges. The weight of the edge defines the probability of a topic given a user.

5   Experimental Setup

The goal of our experiments is to study the co-evolution of social and content networks in social media such as Twitter and Boards.ie and to reveal potential influence patterns between them. We address this by using multilevel time series regression models and analyze bi-directional influences between social and content network properties. Our approach of detecting influences between social network and content network properties requires the following four phases:

5.1   Network Generation

For studying the co-evolution of users' social and content-related behavior, we first need to construct the social and content network from our dataset. Since we have time series data over 30 different time points for the Twitter datasets and twelve different time points for the Boards.ie dataset, we created a social and content network for each specific time point. The social network is a one-mode directed network, where each vertex represents a user and the edges between these vertices represent the directed follow-relations between two users at a certain point in time. To characterize the generated social network of each user at each point in time, we compute the social network properties described in section 5.3.

The content network at each point in time is a two-mode network, which connects users and content-items (e.g., hashtags, keywords, topics) via authoring-relations. To characterize these content networks we compute various content network properties described in section 5.3. It would also be possible to create further types of content networks (e.g., hashtag co-occurrence network) by folding the two-mode content networks which we currently use (see [20] for further types). But we leave the investigation of such network types open for future research.

Finally, we can connect both networks via their user vertexes, since we know which user in the social network corresponds to which user in the content network and vice versa.

5.2   Data Preparation

To prepare the data for the final regression model, some data preparation steps are required. First, the available feature data is normalized by subtracting the mean and dividing the result by the standard deviation which is a common method in applied regression [7]. The subtraction of the mean improves the interpretation of main effects, and the division by the standard deviation brings all values to a common scale. This rescaling helps to interpret the regression coefficients more directly. The fixed effects can be analyzed as the effect of one standard deviation of change in the independent variable on the number of standard deviations change in the dependent variable [22]. Since the final model is autoregressive (see section 5.4) we also have to relate a value at time t with all values at time $t - 1$.

Since our datasets differ substantially, distinct properties are required to capture their characteristics. Note that the centrality measure and clustering coefficient were only calculated for the community dataset since the corresponding social networks of our randomly-sampled and Boards.ie datasets also includes edges to users outside the group of seed users, so calculating such measures would not make sense.

5.3   Network Properties

Because we are interested in how content and social networks evolve over time and which factors impact the evolution, we need a set of properties which characterize our networks. Those properties will help us to track how the networks change over time and how different properties of the network influence others.

Social Network Properties

Degree Measures. The *degree* $d(v)$ is the number of edges of a vertex v which is equal to the number of linked neighbors of the vertex. In directed networks, the *in-degree* of a vertex represents the number of incoming edges and the *out-degree* represents the number of outgoing edges. The *degree centrality* is equal to the degree of a vertex v divided by the maximum possible degree represented by the number of other vertices in the network. Again one can measure *in-degree centrality* and *out-degree centrality* in directed networks. Equation 2 defines the degree centrality, which indicates which actor is most active (or central) by counting the number of ties the actor has to other actors in the network or graph [23].

$$C_D(v) = \frac{d(v)}{n - 1} \qquad (2)$$

Betweenness Centrality. The *betweenness centrality* of a vertex v is the fraction of geodesics (shortest paths) between other vertices that run through v over all geodesics in the whole network. The main idea behind the betweenness centrality is that an actor is central if he/she lies between other actors on their geodesics [23]. The betweenness centrality may also help to quantify how influential a specific person is in the network [8]. The betweenness centrality is defined as follows, where σ_{st} defines the number of geodesics from s to t ($s, t \in V$) and $\sigma_{st}(v)$ refers to the number of shortest paths from s to t that go through v.

$$C_B(v) = \sum_{s \neq v \neq t \in V} \frac{\sigma_{st}(v)}{\sigma_{st}} \tag{3}$$

Clustering Coefficient. Watts and Strogatz [24] define the *clustering coefficient* C_v as the number of actual links between the neighbors of a vertex v divided by the number of possible links between the neighbors of v. Since the number of possible edges between the neighbors is different for directed and undirected graphs, the definition of the clustering coefficient is as well different for directed and undirected networks. Equation 4 shows the clustering coefficient for directed networks because we only use directed networks in this work.

$$C_i = \frac{|\{e_{jk}\}|}{k_i(k_i - 1)} : v_j, v_k \in N_i, e_{jk} \in E. \tag{4}$$

Content Network Properties

Number of Tweets. The *number of tweets* property defines how many tweets a user (represented via a vertex v) authored at a certain point in time point t and is defined as follows:

$$NT(v,t) = \#Tweets(v,t) \tag{5}$$

Hashtag Ratio. The *hashtag ratio* represents the number of hashtags used by a user (represented via a vertex v) per day t, normalized by the number of daily tweets authored by him ($NT(v,t)$). We define the hashtag ratio as follows:

$$R_h(v,t) = \frac{\#Hashtags(v,t)}{NT(v,t)} \tag{6}$$

Link Ratio. The *link ratio* represents the number of daily used URLs by a user v, normalized by the number of tweets he/she published that day and is defined as follows:

$$R_l(v,t) = \frac{\#Links(v,t)}{NT(v,t)} \tag{7}$$

Retweet Ratio. The *retweet ratio* represents how often a user v has retweeted messages published by other Twitter users, normalized by the number of tweets he/she published that day. We define the retweet ratio as follows:

$$R_{rt}(v,t) = \frac{\#Retweets(v,t)}{NT(v,t)} \tag{8}$$

Retweeted Ratio. The *retweeted ratio* represents how often other users retweet messages produced by a user v, normalized by the number of tweets user v published that day. The retweeted ratio is defined as follows:

$$R_{rtd}(v,t) = \frac{\#RetweetedTweets(v,t)}{NT(v,t)} \tag{9}$$

Topic Entropy. The *topic entropy* measures the concentration of a user v across a bunch of topics z which are discussed on Boards.ie. The topic entropy of a user helps us to gauge the topical authoring behavior of a user based on the topic distribution learned from the aggregation of posts he/she authored within the last six month. A user can either have a broad repertoire of topics throughout his/her posts, or can focus his/her authoring behavior on few selected topics. A higher entropy of $\hat{\theta}_{v_i}$ indicates a more random topic distribution - suggesting that a user talks about many different topics - while a lower entropy value corresponds to the user being concentrated on a lower number of topics. We define the topic entropy of a user v at time t as follows:

$$H_Z(v_t) = -\sum_z p(z|v_t) \log p(z|v_t) \tag{10}$$

5.4   Analyzing Influence Patterns

Based on the prepared data, the final model described in section 3 can be applied to identify a potential influence between social and content network properties at different points in time. The dependent variable is always a property at time t and the independent variable are all properties at time $t - 1$ including the dependent variable at that time. Including the dependent variable in that step allows us to detect if a variable's previous value influences it's future value. Finally, the resulting statistical significant coefficients show a relationship between an independent variable at time $t - 1$ and a dependent variable at time t. Positive coefficients indicate that a rising value of a property leads to an increase of another property, while negative coefficients indicate that a rising value of a property leads to an decrease of another property. To reveal positive and negative influence relations between properties within and across different networks, we visualize the resulting coefficients of our fitted model as arrows in graphical influence networks. Note that only statistically significant coefficients are illustrated in these graphical influence networks. In order to determine if a coefficient is statistically significant we calculated if the estimate is larger than two times the standard error. If the estimate would be equal or smaller than the standard error, it could not be guaranteed that it is a statistical significant influence, because the standard error could falsify the estimate result. If the estimate is now at least two times larger then it can be stated that the influence is statistical significant. This is a very simple, but useful technique to determine statistical significance. There are many more mechanics available like the t-test. We have compared our significance results to the ones of a t-test and have observed that the results are identical.

6   Results

Our results reveal interesting influence patterns between social networks and content networks of our datasets. Our experiments focus on our two Twitter datasets as well as on our Boards.ie forum dataset. Figure 1 shows the influence network which we obtained by performing a multilevel regression analysis on our randomly-sampled Twitter dataset (see section 4.1), while figure 2 illustrates the results obtained by conducting the same analysis on our community Twitter dataset (see section 4.2). The influence network obtained from our Boards.ie dataset is illustrated in figure 3. The influence networks obtained from these datasets show the correlations detected in the multilevel regression analysis via arrows that point out influences between a property at time t on another property at time $t + 1$. An arrow between two properties indicates that the value of one property at some time point t has a positive or negative effect on the value of the other property at time $t + 1$. How the arrows are calculated is described in section 5.4. Red dashed arrows in figures 1, 2 and 3 represent negative effects and blue solid arrows illustrate positive effects. The thickness of the lines indicates the weight of the influence relations. Only statistically significant influences are showed.

This section is now split into two sub-sections. Section 6.1 discusses results obtained from our randomly-sampled and community-centered Twitter datasets, whereas section subsec:boardsieresults presents results observed from the experiments on the Boards.ie dataset.

6.1   Twitter Results

Interestingly, the influence networks from both datasets (see figures 1 and 2) reveal significant influences of social properties on content network properties but not vice versa. Our results from both Twitter datasets show that the in-degree has the strongest positive effect on content network properties and that the in-degree influences the link and retweeted ratio. This indicates that users start providing more links in their tweets if their number of followers increases. Not surprisingly, users' tweets are also more likely to get retweeted if their number of followers increases, because more users are potentially reading the users' tweets.

The influence network of our randomly-sampled dataset also points out that the in-degree has positive effects on the number of tweets a user authors the next day, while the influence network of our community dataset shows that users are more likely to retweet more messages the next day, if their number of followers grows. This indicates that in both datasets the amount of content a user publishes rises with increasing number of followers.

Further, figure 1 shows that the out-degree of the social network has positive and negative influences on the content network in our randomly-sampled dataset but not in the community dataset. While the positive effects point to the link and hashtag ratio, the negative effects point to the number of tweets and the retweeted ratio. This suggests that users who start following other users also start using more hashtags and links. One possible explanation for this is that users get influenced by the links and hashtags used by users they follow and they might use them more often in their own tweets. The negative effect of out-degree on the number of tweets and the retweeted ratio suggests that users who start following many other users start behaving more like passive readers rather than active content providers.

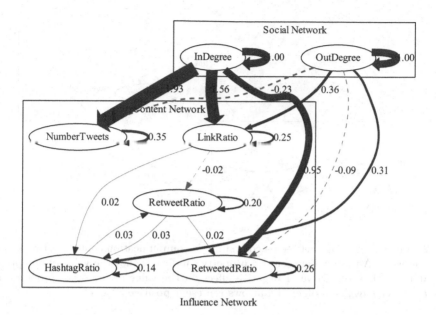

Fig. 1. Influence network between the content and social network of a randomly chosen set of Twitter users. An arrow between two properties indicates that the value of one property at time t has a positive or negative effect on the value of the other property at time $t+1$. Red dashed arrows represent negative effects and blue solid arrows represent positive effects.

On both datasets, all properties influence themselves positively, which indicates that users who are active one day, tend to be even more active the next day. This shows us for example, that users who attract new followers one day tend to attract more new followers the day after or that users who use many hashtags one day tend to use more hashtags the day after. The self-edges for the social properties may be explained by the "rich-get-richer" mechanism. Especially in our first Twitter dataset we are biased towards active users, which may be an indication for this self-influence patterns. Anyway, this does not mean that the number of followers of a user or the number of hashtags a user uses always increases, but that it tends to increase depending on how much it increased the day before. However, the limited period of observation (30 days) in our analysis prohibits us from explaining or investigating this further.

Summary of Twitter Results: Our findings on two small Twitter datasets suggest that there are manifold sources of influence between social and content network properties of our datasets. Our results indicate that users' behavior and the co-evolution of content and social networks on Twitter is driven by social factors rather than content factors. Further work is required to confirm or refute this observations on other, larger datasets. Our experiments also suggest that in-degree strongly influences properties of the content network, which we interpret

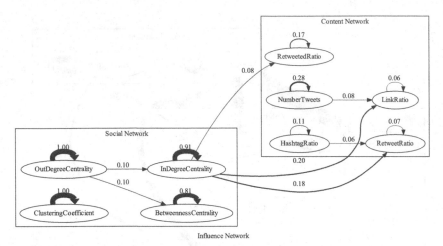

Fig. 2. Influence network between a community's communication content network and social network. An arrow between two properties indicates that the value of one property at time t has a positive or negative effect on the value of the other property at time $t + 1$. The available blue solid arrows represent positive effects.

to mean that the number of followers is a very important motivation for Twitter users to add more content and use more content features like hashtags, URLs or retweets.

6.2   Boards.ie Results

One can see from figure 3 that content properties also weakly influence social properties on Boards.ie, which was not the case for both Twitter datasets. This indicates that users' reply behavior (and therefore their social network of communication partners) is also influenced by content properties. But one needs to take into account that these effects are very weak and may depend on the way how the social network was constructed - it was induced from the content a user produces.

Like in our Twitter experiments all social properties influence themselves. The strong positive influence of the out-degree on itself means that the number of users to which a user replies increases depending on to how many users he/she replied the day before. The influence of the in-degree on itself is not as strong, but it also implies that the number of replies a user gets one day increases depending on how much replies he/she got the day before. This indicates that often discussions start on Boards.ie and therefore users who are involved in these discussions start increasing their communicating network via their reply-activities and the reply-reactions to others.

The only content network property, the topic entropy, also shows a positive impact on itself. It suggests that a user tends to extent his topical focus, depending on how broad his topical focus was the month before.

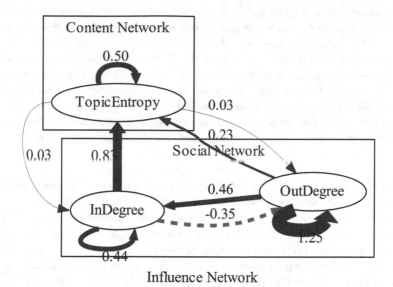

Fig. 3. Influence network between the content network and communication-partner network of the Boards.ie dataset. An arrow between two properties indicates that the value of one property at time t has a positive or negative effect on the value of the other property at time $t + 1$. Red dashed arrows represent negative effects and blue solid arrows represent positive effects.

The in-degree shows a positive influence on the topic entropy, which means that a user who gets many replies is extending his/her topical focus the next month during our observation period. The same effect can be observed for the out-degree, which affects the topic entropy positively. This means that a user who replies to many others is extending the topical focus the next month. One possible explanation for that is that users get influenced by the topics of their communication partners, amongst others because discussions on Boards.ie can be lengthy and users might often adapt topics of their communication partner by answering to their replies.

Summary of Boards.ie Results: This experiment shows us that users who are active in Boards.ie tend to get more involved into the Boards.ie forum over time, because they seem to post more replies and also get more replies from others. Furthermore, social reply behavior has an effect on the topical behavior of users.

7   Conclusions, Limitations and Future Work

The main contributions of this paper are the following: (i) We applied multilevel time series regression models to two substantially distinct Twitter datasets and one dataset from Boards.ie consisting of a temporal series of social and content data and (ii) we explored and identified influence patterns between social and content networks on Twitter and Boards.ie.

In our experiments, we studied how the properties of social and content networks evolve over time, and we explored influence patterns between them. We showed that the adopted approach allows tackling interesting questions about how communities on Twitter evolve and the factors that impact communities' content-related and social behavior over time. Furthermore, this approach also allows us to detect the influence on topic and reply behavior of social media users based on a public forum called Boards.ie.

Our analyses may also facilitate community managers to promote certain features of the platform and steer communities and their behavior. For example, one can see from our analysis that usage of content features, such as hashtags and links, is highly influenced by social network properties such as in-degree. Therefore, community managers could try to encourage users to use more content features by introducing new measures such as a friend recommender techniques which might impact the social network of users. Based on our analysis on the public message board dataset, forum managers might also gain information about what influences the topical authoring behavior of users - i.e. what makes them focusing on few topics rather than authoring posts about a broad variety of topics. However, further work is warranted to study these ideas.

In addition, our results yield a number of interesting observations about the nature of interactions between social and content networks on platforms such as Twitter or Boards.ie, including the following:

- *The growth of a Twitter user's followers increases the number of links they use per tweet and the number of times their tweets get retweeted,*
- *An increase of a Twitter user's followers increases the number of tweets that get retweeted on average,*
- *The number of hashtags or links Twitter users use per tweet does not impact their number of followers and*
- *Users of the online forum Boards.ie get more involved into the board over time and broaden the topical focus of their authoring behavior.*

Another interesting hypothesis emerging from our work is that the co-evolution of content and social networks on different social media applications is driven by distinct factors. For example, our results suggest that the evolution of social networks in our Twitter datasets strongly influences the evolution of content network in these datasets *but not vice versa*. This would mean that Twitter users' behavior and consequently the co-evolution of social and content networks on Twitter is driven by social factors rather than content factors. However, for example on Boards.ie we observed that not only social networks influence the evolution of content networks, but also content networks impact the evolution of social networks (to a lesser extent). This is not surprising since we had to induce the social network from content for our Boards.ie dataset, because no explicit social network exists. Related research by Anagnostopoulos et al. [2] found that content on Flickr is not strongly influenced by social factors. In the light of our investigations, this may suggest that different social media applications and different communities may be driven by different factors. The experimental setup used in our work can be applied to different datasets to study these questions in future.

Our results from all datasets show that the in-degree of the social network has a strong effect on the content network, which indicates that the attention of other users is a very important motivation for social media users to add more content and use more content features like hashtags, URLs or retweets or broaden their topical focus. The out-degree, that means the number of users a user is following or replying to, can however even have a negative influence on content network properties as one can see in the influence network of the randomly-sampled dataset (see figure 1). It suggests that an increase of a Twitter user's followees (i.e., the number of users he/she follows) implies that the user starts tweeting less and that his/her tweets get less frequently retweeted. Further our findings show that all properties influence themselves positively in all three datasets. This does not mean that the values of all properties always increase over time, but that they tend to increase depending on how much they increased the day before. For example, a Twitter user who started posting more links at day t, is likely to post even more links at day $t + 1$ or a user who gain new followers at day t is likely to gain even more new followers at day $t + 1$.

Based on the techniques developed by Wang and Groth [22], our work investigated influence patterns in a new domain, i.e. on microblogging platforms like Twitter, and applied an approach of topical behavior detection on a public forum like Boards.ie. Our results are relevant for researchers interested in social network analysis, text mining and behavioral user studies, as well as for community manager and hosts which need to understand the factors that influence the evolution of their communities in terms of their content-related and social behavior.

All results presented in this work are limited to the available datasets which consist of a small sample set of Twitter and Boards.ie users. Further we want to point out that our work identifies correlations between properties of content and social networks of our datasets rather than causal effects. By analyzing such relations over time we aim to approximate causality but further studies are required to verify causal effects. Although our work has certain limitations, it illuminates a path towards studying complex dynamics of social and content networks' evolution on social media systems such as Twitter or Boards.ie.

Acknowledgments. This work is in part funded by the FWF Austrian Science Fund Grant I677 and the Know-Center Graz. Claudia Wagner is a recipient of a DOC-fForte fellowship of the Austrian Academy of Science. The authors want to thank Boards.ie for making their data available and Matthew Rowe for constructing the social network of Boards.ie users.

References

1. Aiello, L.M., Barrat, A., Cattuto, C., Ruffo, G., Schifanella, R.: Link creation and profile alignment in the aNobii social network. In: SocialCom 2010: Proceedings of the Second IEEE International Conference on Social Computing, Minneapolis, Minnesota, USA, pp. 249–256 (August 2010)

2. Anagnostopoulos, A., Kumar, R., Mahdian, M.: Influence and correlation in social networks. In: Li, Y., Liu, B., Sarawagi, S. (eds.) Proceedings of the 14th ACM SIGKDD International Conference on Knowledge Discovery and Data Mining, Las Vegas, Nevada, USA, August 24-27, pp. 7–15. ACM (2008)
3. Bakshy, E., Hofman, J.M., Mason, W.A., Watts, D.J.: Everyone's an influencer: quantifying influence on twitter. In: Proceedings of the Fourth ACM International Conference on Web Search and Data Mining, WSDM 2011, pp. 65–74. ACM, New York (2011)
4. Blei, D.M., Ng, A., Jordan, M.: Latent dirichlet allocation. JMLR 3, 993–1022 (2003)
5. Cha, M., Haddadi, H., Benevenuto, F., Gummadi, P.K.: Measuring user influence in twitter: The million follower fallacy. In: Cohen, W.W., Gosling, S. (eds.) Proceedings of the Fourth International Conference on Weblogs and Social Media, ICWSM 2010, Washington, DC, USA, May 23-26. The AAAI Press (2010)
6. Crandall, D., Cosley, D., Huttenlocher, D., Kleinberg, J., Suri, S.: Feedback effects between similarity and social influence in online communities. In: Proceedings of the 14th ACM SIGKDD International Conference on Knowledge Discovery and Data Mining, KDD 2008, pp. 160–168. ACM, New York (2008)
7. Gelman, A.: Scaling regression inputs by dividing by two standard deviations. Statistics in Medicine 27(15), 2865–2873 (2008)
8. Goh, K.-I., Oh, E., Jeong, H., Kahng, B., Kim, D.: Classification of scale-free networks. Proceedings of the National Academy of Sciences of the United States of America 99(20), 12583–12588 (2002)
9. Griffiths, T.L., Steyvers, M.: Finding scientific topics. PNAS 101(suppl. 1), 5228–5235 (2004)
10. Hayes, A.F.: A primer on multilevel modeling. Human Communication Research 32(4), 385–410 (2006)
11. Kitagawa, G.: Introduction to Time Series Modeling (Chapman & Hall/CRC Monographs on Statistics & Applied Probability). Chapman and Hall/CRC (2010)
12. Krner, C., Grahsl, H.-P., Kern, R., Strohmaier, M.: Of categorizers and describers: An evaluation of quantitative measures for tagging motivation. In: 21st ACM SIGWEB Conference on Hypertext and Hypermedia (HT 2010). ACM (2010)
13. Kwak, H., Lee, C., Park, H., Moon, S.: What is Twitter, a social network or a news media? In: WWW 2010: Proceedings of the 19th International Conference on World Wide Web, pp. 591–600. ACM, New York (2010)
14. McCallum, A.K.: Mallet: A machine learning for language toolkit (2002), http://mallet.cs.umass.edu
15. Naveed, N., Gottron, T., Kunegis, J., Alhadi, A.C.: Bad news travel fast: A content-based analysis of interestingness on twitter. In: WebSci 2011: Proceedings of the 3rd International Conference on Web Science (2011)
16. Rowe, M.: Forecasting audience increase on youtube. In: User Profile Data on the Social Semantic Web Workshop, Extended Semantic Web Conference (2011)
17. Singer, P., Wagner, C., Strohmaier, M.: Understanding co-evolution of social and content networks on twitter. Making Sense of Microposts 4 (2012)
18. Skrondal, A., Rabe-Hesketh, S.: Generalized Latent Variable Modeling: Multilevel, Longitudinal, and Structural Equation Models. Chapman and Hall/CRC (2004)

19. Suh, B., Hong, L., Pirolli, P., Chi, E.H.: Want to be retweeted? large scale analytics on factors impacting retweet in twitter network. In: Elmagarmid, A.K., Agrawal, D. (eds.) Proceedings of the 2010 IEEE Second International Conference on Social Computing, SocialCom / IEEE International Conference on Privacy, Security, Risk and Trust, PASSAT 2010, Minneapolis, Minnesota, USA, August 20-22, pp. 177–184. IEEE Computer Society (2010)

20. Wagner, C., Strohmaier, M.: The wisdom in tweetonomies: Acquiring latent conceptual structures from social awareness streams. In: Proc. of the Semantic Search 2010 Workshop (SemSearch 2010) (April 2010)

21. Wallach, H.M., Mimno, D., McCallum, A.: Rethinking LDA: Why priors matter. In: Proceedings of NIPS (2009)

22. Wang, S., Groth, P.: Measuring the Dynamic Bi-directional Influence between Content and Social Networks. In: Patel-Schneider, P.F., Pan, Y., Hitzler, P., Mika, P., Zhang, L., Pan, J.Z., Horrocks, I., Glimm, B. (eds.) ISWC 2010, Part I. LNCS, vol. 6496, pp. 814–829. Springer, Heidelberg (2010)

23. Wasserman, S., Faust, K.: Social Network Analysis: Methods and Applications. Structural Analysis in the Social Sciences. Cambridge University Press (1994)

24. Watts, D.J., Strogatz, S.H.: Collective dynamics of 'small-world' networks. Nature 393(6684), 440–442 (1998)

25. Zubiaga, A., Körner, C., Strohmaier, M.: Tags vs shelves: from social tagging to social classification. In: Proceedings of the 22nd ACM Conference on Hypertext and Hypermedia, HT 2011, pp. 93–102. ACM, New York (2011)

Mining Dense Structures to Uncover Anomalous Behaviour in Financial Network Data*

Ursula Redmond, Martin Harrigan, and Pádraig Cunningham

School of Computer Science & Informatics,
University College, Dublin
ursula.redmond@ucdconnect.ie, {martin.harrigan,padraig.cunningham}@ucd.ie

Abstract. The identification of anomalous user behaviour is important in a number of application areas, since it may be indicative of fraudulent activity. In the work presented here, the focus is on the identification and subsequent investigation of suspicious interactions in a network of financial transactions. A network is constructed from data from a peer-to-peer lending system, with links between members representing the initiation of loans. The network is time-sliced to facilitate temporal analysis. Anomalous network structure is sought in the time-sliced network, illustrating the occurrences of unusual behaviour among members. In order to assess the significance of the dense structures returned the enrichment of member attributes within these structures is examined. It seems that dense structures are associated with geographic regions.

1 Introduction

This work is part of a project whose purpose is to locate anomalous user behaviour which may represent fraud within a financial transaction system. The focus is on the detection of groups collaborating with a view to committing fraudulent acts, rather than individuals. Since only a small fraction of such collectives are genuinely behaving maliciously, they remain difficult to isolate.

With this goal in mind, we analyze financial transaction data from an online peer-to-peer lending website. In this setting, regulation is difficult since loans are unsecured and borrowers have not undergone the rigorous examinations required by traditional banks. Thus, the scope for committing acts of fraud is quite broad, especially since these peer-to-peer systems are still reasonably new.

This paper aims to employ network analysis methods to uncover suspicious groups of members in such a peer-to-peer lending system. Since transactions create a relation between users, a network structure captures the data well. The detection of interacting parties through the use of graph theoretic methods is enabled by this choice of data representation. A brief introduction to some necessary vocabulary follows.

A graph is a mathematical abstraction of a network. A graph G consists of a set of vertices V and a set E of pairs of vertices called edges. If these vertex

---

* This work is supported by Science Foundation Ireland under Grant No. 08/SRC/I1407.

M. Atzmueller et al. (Eds.): MSM/MUSE 2011, LNAI 7472, pp. 60–76, 2012.

pairs are ordered, the graph is directed. An edge is directed from a source vertex to a target vertex. A graph may be symbolically represented as $G = (V, E)$. The graph $H = (W, F)$ is a subgraph of G if W is a subset of V and F is a subset of E, such that the source and target vertices of each edge in F are in W. In a directed graph, the number of edges with a vertex v as source is the out-degree of v, and the number of edges with v as target is the in-degree of v.

The problem of finding a subgraph H embedded in a graph G is known as the subgraph isomorphism problem. A graph H is subgraph isomorphic to a graph G if there exists a one-to-one correspondence between the vertex sets of H and a subgraph of G, and adjacency between corresponding vertices is preserved. This is known to be an NP-complete problem [1]. Preprocessing a network to identify relevant structures apriori allows for subsequent fast access to them for the user, although conceding the cost of an increase in storage space. This approach aims to reduce as much as possible the necessity to perform subgraph isomorphism testing and motivates the approach taken in this paper.

The layout of this paper is as follows. Section 2 reviews related work from a number of fields. Section 3 describes peer-to-peer lending systems, which provide the data to be analyzed. Section 4 describes the network extracted from the peer-to-peer lending data. Section 5 presents some preliminary results. Among these is a description of how certain structures are distributed, along with an analysis of the actors involved. Section 6 concludes the paper and suggests future work.

2   Related Work

In quite a diverse range of networks, fraud is a problem. To combat this, network analysis methods have been developed to detect fraudulent behaviour. Since criminals groups tend to form communities, the discovery of closely connected individuals may lead to their identification. This task calls for the mining of dense structures in the network. Another form of collaboration among fraudsters is exhibited as connected collections of paths through a network. For these interactions to be associated with a single incident, the time at which they occurred is important. Thus, a temporal analysis of the network is required. In order to avoid as much as possible the requirement of subgraph isomorphism testing, a popular approach has been to seek frequent subgraphs, which by construction cover more of the graph than infrequent subgraphs. An index of such subgraphs allows for more efficient subgraph search in the graph overall, and motivates the direction of our current work and the course it will take in the future.

2.1   Fraud Detection in Network Data

The tools of social network analysis have proven effective in the search for malicious behaviour in a variety of contexts. For example, to detect automobile insurance fraud [2], relations between drivers, incidents and cars were examined in a network context. Given this data representation, fraud was efficiently uncovered, although the networks analyzed were on a small scale.

Healthcare fraud was the focus of a study in which clinical pathways were represented as small temporal graphs [3]. The occurrences of each clinical procedure were ordered in time, with the order represented by directed edges. Frequency mining was performed on the graphs, with frequent graphs isolated as features. A classification step allowed an example graph to be recognized as fraudulent or legitimate.

To detect the spread of misinformation in the context of political elections in the U.S.A, analysis of networks constructed from social media was performed [4]. Specifically, a data set of memes were examined, and their sentiment was analyzed. A web interface was used to present statistical information and an interactive visualization of the data. An interactive annotated timeline allowed temporal data to be explored.

A criminal network composed of terrorists was examined from the point of view of the links between members [5]. A denser network led to ease of communication, and correspondingly to ease of detection. A sparser network delayed communication, but made the organization more difficult to detect. Hence, the trade-off facing criminals is between the efficiency and security of their interactions.

Fraudulent behaviour in financial networks often exhibits signature structure [6]. For example, *smurfing* is the splitting of a large financial transaction into multiple smaller transactions, each of which is below a limit which, if exceeded, would cause the given financial institutions to take note. This behaviour can be represented by a fan-out-fan-in (FOFI) structure. This captures the movement of funds from a source to a destination, with the amount divided between an intermediate set of accounts, in order to keep each transaction under the threshold.

In a telecommunications network, the accounts of fraudsters tend to cluster together [7]. This suggests a near-clique representation for fraudulent activity. In the case of an online auction community, fraudsters may interact in small and densely connected near-cliques with the aim of mutually boosting their credibility [8]. Once an act of fraud is committed, the near-clique can be identified by the auction site and the accounts of fraudsters removed. To exploit the system again, the group must reconstruct their network from scratch.

2.2   Dense Structure Mining

In social networks, dense network regions indicate the existence of a community. Fortunato [9] presents a comprehensive review of the broad field of community finding. This includes the many subtly different definitions of community structure, and the methods used to find them, including those of a statistical and graph theoretical leaning.

The densest type of network structure is a clique, in which each vertex connects to every other vertex in the clique. An algorithm devised by Bron and Kerbosch [10] was employed to find all maximal cliques (cliques that are not contained in any larger clique). There are a number of relaxations of the notion of a clique. A k-plex [11] is a maximal subgraph with all vertices adjacent to at least all but k of the members. A k-core [12] is a maximal subgraph with all vertices adjacent to at least k of the members.

A k-truss [13,14] strikes a good balance between restrictiveness and tractability. A k-truss is a non-trivial connected subgraph in which every edge is part of at least $k-2$ triangles within the same subgraph. A maximal k-truss is a k-truss that is not a proper subgraph of another k-truss. Cohen [13] shows that a clique with k vertices contains a k-truss. However, a k-truss need not contain a clique with k vertices. On the tractability side, if we assume a hash table with constant time lookups, then for a graph $G = (V, E)$, all maximal k-trusses can be enumerated in $\mathcal{O}\sum_{u \in V} deg(u)^2$-time, where $deg(u)$ is the degree of a node u. The running time can be reduced in practice by using the maximal $k-1$-trusses of G as input.

A graph may be reduced to a k-truss, examples of which are illustrated in Figures 8 and 7, via an algorithm by Cohen [14]. To greatly speed the computation, the maximal $k-1$-core of the graph is first located, and the elements not included in this structure are removed. Then, edges lacking the required number of supporting edge pairs are removed. Finally, isolated vertices resulting from previous steps are removed.

A network may be analyzed by examining the sizes and distributions of various types of dense structure present. Outliers in these distributions may also be identified and examined. For example, instances of particularly dense structures that are formed in a relatively short period of time may be of particular interest.

2.3   Temporal Network Analysis

Holme and Saramäki [15] review a number of topological structures, methods and models that are applicable in a temporal context. For example, in directed graphs, each edge has a corresponding timestamp. Time-respecting paths are described as paths with the further constraint that they are sequences of link activations that follow one another in time. In a traversal of the edges of a time-respecting path, no edge has a timestamp earlier than the timestamp of the edge preceding it. This definition leads to time-respecting components and hence, to temporal variations in network structure. In this paper, time-sliced graphs are used to ensure that paths are time-respecting.

In the context of financial transactions, such paths define the set of members through whom funds may pass during a specified time. A FOFI structure, as illustrated in Figure 9, has a number of paths and hops between a source and target vertex. The appearance of FOFI and other types of subgraph may be bursty (occur in large quantities during short time intervals). An analysis of the temporal existence of such structures may give support to the claim that they are suspicious. Thus, we let the appearance of connected time-respecting paths during isolated time intervals guide our further analysis.

2.4   Frequent Structure Mining and Index Construction

Vanetik *et al.* extended the Apriori algorithm to identify frequent subgraphs [16]. The subgraphs are treated as sets of paths, and the Apriori join operates on these paths. The size of the maximal independent set of instances of a subgraph is used as an admissible support measure, where instances must be edge-disjoint. GREW [17] is a heuristic algorithm which finds connected subgraphs with a large number of node-disjoint embeddings. The algorithms HSIGRAM and VSIGRAM [18] discover frequent subgraphs using breadth-first and depth-first search paradigms respectively. Connected subgraphs with a specified number of edge-disjoint embeddings are found in an undirected labeled sparse graph.

An index is an effective way to store mined structures. The gIndex algorithm [19] mines frequent subgraphs in a database of graphs and uses these as an index. The FG-index [20] proposes a nested inverted-index, based on the set of frequent subgraphs. If a graph query is a frequent subgraph in the database, this index returns the exact set of query answers, with no need for candidate verification. If the query is an infrequent subgraph, a candidate answer set close to the exact answer set is produced. In this case, the number of subgraph isomorphism tests will be small, since, by construction, an infrequent subgraph will appear in few graphs in the database.

3   Peer-to-Peer Lending Systems

Peer-to-peer lending and crowdfunding have emerged in recent years as the financial dimension of the social web. They present an interesting application area due to the open availability of data[1]. Initiatives in these areas offer the potential benefit of disintermediating the process of fundraising and borrowing money. Example peer-to-peer lending systems are prosper.com, lendingclub.com and zopa.com. Closely related to peer-to-peer lending is the idea of crowdfunding or crowdsourced investment funding. Two typical crowdfunding companies for startup funding are seedups.com and crowdcube.com There have also been a number of initiatives around crowdfunding for creative projects (e.g. poziblie.co.uk). However, these initiatives are of less relevance here as they are altruistic rather than investment initiatives.

The risk of misuse and fraud is at least as great in these new online systems as in traditional bricks-and-mortar finance. It is important to monitor the networks of transactions on these sites to check for money laundering and fraud. This is important for reasons of regulation but also to encourage users to trust the service. Indeed for this reason prosper.com, a prominent peer-to-peer lending system, make all of their transaction data available for scrutiny.

3.1   The Prosper Lending System

Prosper opened to the public in February 1996. It was closed briefly during 2008 and again during 2009 due to regulatory issues. However, it has reported significant growth in recent months[2]. As of September 2011, there was a record of 8 916 105 bids on 401 180 listings between 1 207 418 members. Of these listings, 43 576 were accepted as loans. There were 4 019 approved groups organized into 1 566 categories. Borrowers join groups in order the improve their chances of

---

[1] http://www.prosper.com/tools
[2] WSJ.com - Peer-to-Peer Loans Grow http://on.wsj.com/kFUAuy

having their listings funded; many groups enforce additional identity checks on their members and garner a certain amount of trust.

The management of trust and reputation is an important feature in such peer-to-peer systems. Information asymmetry (the difference in the quantity of information a lender and borrower have about each other) can have a substantial effect on the behaviour of users [21]. There is a reasonably close-knit community of lenders on the Prosper website, who advise each other with respect to lending strategies. For example, analysis of individual lender and borrower activity is summarized by blog posts and in forums on the Internet[3]. Members who behave in ways that have proven unsuccessful are named as examples not to follow.

3.2   The Prosper Data Set

The Prosper website provides a nightly snapshot of all data pertaining to listings, bids, users, groups and loans. These data are released in order to facilitate the statistical analysis of the system as it grows and develops. The version providing data for the analysis in this work comes from the time period spanning November 2005 to September 2011.

A member of the Prosper website is a registered user, who may have one or multiple roles, including that of borrower, lender, group leader or trader. The city and state of residence of members is available, along with a list of their friends on the website, any endorsements they have been given, a description of themselves and a picture.

A listing is created by a borrower in order to solicit bids. If enough are received to reach the amount requested, then after the listing period ends, the listing becomes a loan. A bid is created by a lender, specifying an amount and a minimum rate required, should the bid win the auction and become a loan. A loan is created when a borrower has enough bids to meet their requested amount. The possible statuses of a loan include current, late, paid, defaulted upon and cancelled.

A group accumulates members who have similar interests or affiliations. Group members who are borrowers often get better interest rates, since lending group members have more confidence in those who belong to trusted groups. Groups are categorized according to similarity of interests.

4   Network Description

A network was extracted from the Prosper data set to facilitate the detection of suspicious structure. Features of the network as a whole, and of its time-sliced components, are presented in this section. The Python programming language [22], which provides the NetworkX [23] and matplotlib [24] libraries, were used to generate the results and plots presented in this paper. Network structures were visualized using the Gephi [25] framework.

---

[3] A Different Look at Prosper's Loan Origination Trend http://www.prospers.org/blogs/ira01/2008/08/09/loans_with_very_high_rates_for_their_cre.html

4.1   The Prosper Loan Network

A network composed of Prosper members may be constructed via the listings that are posted by borrowers and bid upon by lenders. Since the flow of money through this network is of interest, only listings which were fully funded and became loans were used to link members.

The vertex set consists of members who received or contributed to loans at some point during their career at Prosper. Only those vertices involved in a transaction were retained (those involved in listings that weren't fully funded are not included). The edge set captures the flow of money between these vertices, from members who acted as lenders to those who acted as borrowers for the purpose of the transaction. The edge is time-stamped with the origination date of the loan, which marks the time when the borrower received funds and amortization began.

The network takes the form of a directed graph with parallel edges, but no self-loops. The number of vertices is 89 269 and the number of edges is 3 454 649. The average degree is 38.699. The density of the graph is 4.335×10^{-4}.

An analysis of the degree distributions is presented in Figure 1. In Figure 1a, a log-log plot of all three degree distributions is shown. As the degree increases, the number of vertices with that degree decreases, but in an increasingly unpredictable way. The cumulative distribution function (CDF) of the three degree distributions is presented in Figure 1b. To compute this, first the probability density function for each distribution is calculated. This scales the sum of the values to between zero and one. The CDF illustrates the probability that for some degree x chosen at random from the range presented, vertices will be found with a degree less than or equal x.

The lower number of vertices with high in-degree could be explained by the limited number of bids required to fully fund a loan. The largest of the loans in this dataset are for $25 000. On the other hand, a member may theoretically place bids on as many loans as they wish. Thus, the out-degree is not so limited.

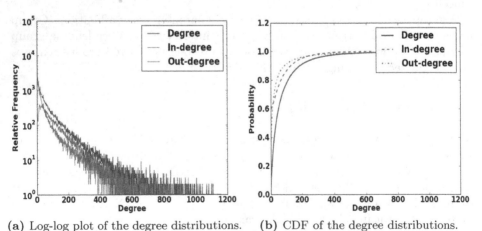

(a) Log-log plot of the degree distributions.   (b) CDF of the degree distributions.

Fig. 1. The degree, in-degree and out-degree distributions of the network

There are 85 746 maximal strongly connected components and 47 maximal weakly connected components. The largest strongly connected component accounts for 3 513 (3.94%) of the vertices, while the giant weakly connected component is composed of 89 171 (99.89%) of the vertices.

The degree-assortativity of a network is a measure of the preference for vertices to attach to others of a similar degree. The value for this network is approximately -0.03, implying slight dissasortativity. Thus, there is no notable tendency for vertices of similar degree to attach to each other.

4.2   Time-Slicing the Loan Network

In order to detect a flow of money along edges within an isolated piece of network structure, temporal analysis comes into play. For example, given three members represented by vertices u, v and w, where u lends to v and v lends to w, it cannot be inferred that money was transferred from u to w if the transactions which occurred were temporally independent of each other. This aspect is captured by the term of each loan, measured in months. Each loan has an origination date, which marks the time when the borrower receives funds and amortization begins. As long as the transaction from v to w begins before the term of the loan from u to v has completed, it is reasonable to claim that money may have been transferred indirectly from u to w.

The minimum term for any loan in the data set is of 12 months duration. Thus, any section of the network in which transactions were carried out within the same 12 months will allow for a flow of money between the members active therein. For the purposes of this paper, time-slices of four months duration were used, with a sliding window of two months. This allows for the amortization of a loan to have occurred, and for the associated funds to be reasonably quickly propagated in the network. Thus, a new slice occurs every two months and is indexed by the first month. Each slice contains loans that began any time from the start of the first month to the end of four months later. This yielded 36 time-slices.

In all of the plots in Figure 2, the temporary closure of the Prosper system during 2008 and 2009 is apparent. The drop-off at the end of each plot is an artifact of how the time-slices were constructed – since each time-slice starts from the indexed month, the final time-slice has only one month-worth of data.

As could be expected, at each opening of the system, the number of active members generally increases, as shown in Figure 2a. The number of connections between users increases correspondingly, as is shown by the edge count in Figure 2b, and the average degree, as in Figure 2c. The density is highest when the system opens, since there are fewer members between which connections can be made. The network becomes more sparse as the number of members increases. The average shortest path length peaks in the years before Prosper is closed, while continuing to decrease in more recent times, as illustrated in Figure 2d.

It is reasonable to expect that active Prosper members will be either borrowers or lenders and not both, since part of the target market for the platform is people who cannot easily obtain a loan through the banking system. People like this, with a low credit grade, would be unlikely to take the role of lender. Based on

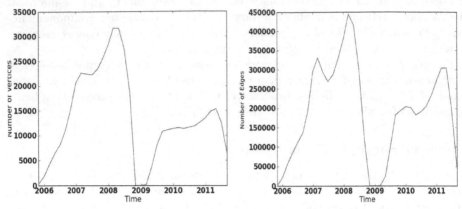

(a) The number of members transacting. (b) The number of monetary transactions.

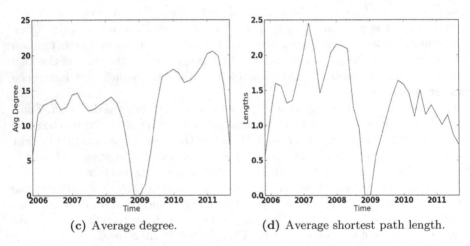

(c) Average degree.   (d) Average shortest path length.

Fig. 2. A temporal analysis of the Prosper network

this assumption, the network structure should be mostly bi-partite. However an examination of the network shows that this is not the case – there are members with incoming and outgoing loans. This behaviour facilitates the existence of densely connected sections of the network, which violate the requirements of a bi-partite network. In these dense regions, the anomalous structures that we seek may be detected. For the purposes of this paper, we limit our search to the discovery of cliques, trusses and FOFI structures where *flows* of money across multiple accounts can occur.

With this in mind, it is interesting to look at the distribution of vertices with both positive in-degree and out-degree. The plot shown in Figure 3 shows that the proportions of lenders who are also borrowers, and the proportions of borrowers who are also lenders, drops after the 2009 shutdown. This may indicate a change in regulation or an uptake in the number of members registering as institutional traders after the shutdown.

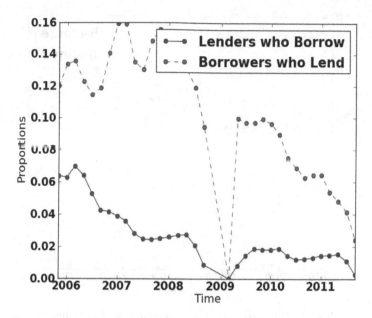

Fig. 3. The proportion of lenders who also borrow, and vice versa

5   Results

In this section we present some details on the distribution of anomalous structures, as their quantities vary throughout the time-sliced network. Some of the larger structures are described in more detail, and we look at the membership of some sample structures to assess enrichment.

5.1   Distribution of Structures

In the version of the Prosper network derived for current study, edges have direction, and there may be multiple edges between vertices. To uncover the structure types sought, the existence of multiple edges between vertices provided little added benefit. Thus, for simplicity, where there were multiple edges, only one was retained. For the algorithms used to detect cliques and trusses, the direction of edges was not important, so graphs were treated as undirected. For FOFI analysis, in which paths define the network structure, graphs were treated as directed.

A plot of the distribution of maximal cliques of size 5 and 6 is shown in Figure 4. As would be expected, there are fewer 6-cliques than 5-cliques in general, although the volume of 5-cliques, especially during mid-2009, is quite striking.

Figure 5 illustrates the distribution of maximal trusses of size 5 and 6. K-trusses are defined by the counts of triangles supporting each edge, so trusses with higher values for k usually have higher edge counts. Rather than enumerating the number of trusses in each time-slice, the number of edges involved in trusses is the statistic we report.

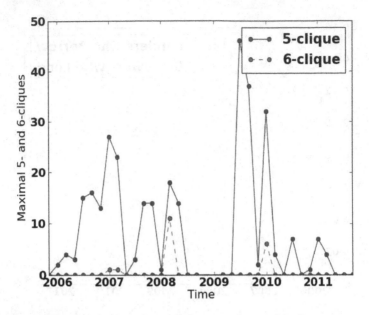

Fig. 4. The distribution of maximal 5- and 6-cliques

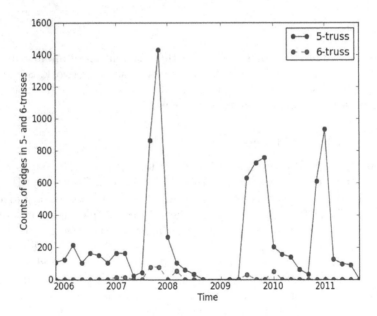

Fig. 5. The distribution of 5- and 6-trusses

Each k-truss is the subgraph of a $k-1$-truss. For example, given a 3-truss in a graph, every edge in the structure must be supported by one triangle. If a 4-truss exists, it is a denser subgraph of this structure, in which every edge is supported by two triangles. This is shown in the plot, since trusses with smaller values for k are always larger; their edges have less strict requirements for support.

FOFI structures may be differentiated by the number of paths and hops they contain. In Figure 6, the distribution of large FOFI structures is shown. As the number of paths increases, the counts of the structure decrease. Interestingly, the existence of FOFI structures of these particular sizes is vastly reduced after Prosper closed in 2008. Again, this may be due to more effective regulation.

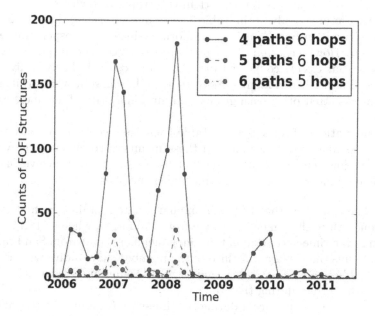

Fig. 6. The distribution of some large FOFI structures

5.2   Structure Analysis

If a set of members, detected through the methods presented in this paper, were to be unexpectedly similar with respect to their attributes, this would provide an external validation that the structure is significant. That is not to say that the structure represents fraudulent behaviour. What is going on may be innocent, but structured flows of funds and connected groups of members acting as lenders and borrowers simultaneously is worthy of investigation. Thus, it is interesting to quantify the extent to which groups or categories are over-represented in the structures detected.

A number of statistical tests are available to quantify this over-representation or enrichment. These tests include Fisher's exact test, Pearson's chi-squared test and the hypogeometric test [26]. In the evaluation that follows, the chi-squared test is used, since some of the structures represent very small portions of the whole graph. It is important to state that statistical significance is not being tested for, as the process entails multiple testing and the p-values have little meaning. Instead, the test statistic ranks all of the dense structures in terms of enrichment. Only structures with low p-values (of less than 10^{-3}) are retained.

The test was applied to the structures deemed most anomalous in terms of size and density. The attributes used were the role, group membership, city and state of the members involved. Since it is not a requirement for all of these fields to be filled by Prosper users, some null values are present. These are ignored, and the most frequent attribute values which are not null are examined. The chi-squared scores are calculated with respect to the time-slice in which the structure of interest occurred, rather than the graph as a whole.

For the structures analyzed in Table 1, it is not surprising that the majority of members reside in San Francisco, California, since the Prosper Marketplace is based there. However, given the diversity of residence over the entire data set, it is notable that closely connected members are extremely likely to live in the same place. Since only 6 of many structures were chosen, somewhat arbitrarily, it may be the case that other such groups contain a majority of members residing elsewhere.

The combinations of roles listed in Table 1 have an over-representation within the structures analyzed. Since most of these members are demonstrably acting as borrowers and lenders simultaneously, it is reasonable to conceive that they have registered more potential roles than other members.

Clique. The time-slice that begins in January 2007 contains the first instance of a 6-clique. All of these members come from California, whereas 18.73% reside there within the time-slice. Half of the clique members come from San Francisco, while 1.88% live there overall. A third of the members have a diverse portfolio of roles, each able to act as borrower, lender, group leader and trader, while 0.55% have all of these roles during that time.

Another set of 6-cliques occur during the time-slice beginning in January 2010. In the one selected, again all members come from California, half come from San Francisco specifically, and a third have roles enabling their actions as borrower, lender, group leader and trader.

Truss. A 5-truss was extracted from the network of November 2006, containing 35 vertices and 104 edges, shown in Figure 8. Two members are in the same group, which has been anonymized as Group A, while 0.14% of members are in that group during that time-slice. This particular truss demonstrates the notion of a cut point, which exaggerates the difference between a clique and a truss. If a truss contains a vertex which acts as a cut point, the removal of this vertex will yield two separate connected components. If any node in a clique is removed, a single connected component remains.

Figure 7 illustrates a 6-truss which occurs in November 2007, and contains 16 vertices and 78 edges. At first glance, this structure has the densely connected appearance of a clique. However, each vertex is not connected to every other vertex. In a 6-truss, each edge being supported by four triangles is the only stipulation.

Table 1. p-values for attribute enrichment

Structure	Attribute	Value	p-value
6-clique, Jan 2007	State	California	1.732×10^{-7}
	City	San Francisco	3.784×10^{-36}
	Roles	Borrower, Lender, Group Leader, Trader	1.391×10^{-27}
6-clique, Jan 2010	State	California	4.650×10^{-7}
	City	San Francisco	2.666×10^{-29}
	Roles	Borrower, Lender, Group Leader, Trader	6.829×10^{-24}
5-truss, Nov 2006	State	California	5.010×10^{-27}
	City	San Francisco	8.996×10^{-62}
	Roles	Borrower, Lender, Group Leader	3.993×10^{-8}
	Group	Group A	5.320×10^{-77}
6-truss, Nov 2007	State	California	1.971×10^{-17}
	City	San Francisco	5.192×10^{-86}
	Roles	Borrower, Lender, Group Leader	8.923×10^{-5}
6-path-5-hop-FOFI, Mar 2006	State	California	1.923×10^{-14}
	City	San Francisco	7.328×10^{-42}
5-path-6-hop-FOFI, Mar 2008	State	California	1.972×10^{-19}
	City	San Francisco	6.776×10^{-87}
	Roles	Borrower, Lender, Group Leader	1.540×10^{-4}

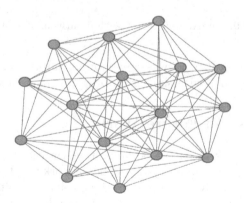

Fig. 7. A 6-truss with 16 vertices and 78 edges occurs in November 2007

FOFI. In Figure 9 is a FOFI structure with 6 paths and 5 hops, which occurs during March 2006. The target vertex receives funds from 6 different sources, all of whom indirectly receive funds from a single source. During March 2008, FOFI structures with 5 paths and 6 hops appear. Details of the enrichment of their attributes are presented in Table 1. It is notable that the number of FOFI

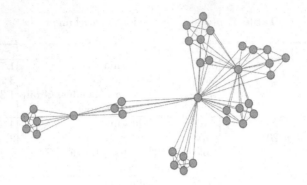

Fig. 8. A 5-truss with 35 vertices and 104 edges occurs in November 2006 and illustrates two cut points

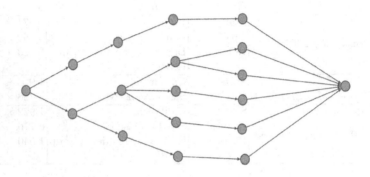

Fig. 9. An example FOFI with 6 paths and 5 hops

structures drops off after Prosper was closed for regulatory reasons. Once Prosper re-opened, only structures with shorter path lengths made a visible come-back.

6   Conclusions and Future Work

In this paper, a network of members interacting via loans was extracted from the Prosper peer-to-peer lending system. The network was time-sliced, in order to facilitate temporal analysis. Since suspicious behaviour often materializes in densely connected parts of a network, and parts with long, interconnected paths, techniques from graph mining were employed to uncover a set of anomalous structures.

The FOFI structure – comprising a source, target and set of intermediate actors – was described, and instances of the structure were detailed. Examples of cliques, and a more relaxed version of a dense structure – the k-truss – were also sought. These structures have been identified, visualized and shown to have a noteworthy distribution of attribute values.

Following on from this work, a more comprehensive method of indexing structures will be developed. In order to exploit the suspicious nature of anomalous structure, the frequency of their occurrence will be taken into account. A visualization system will be developed, so as to provide an intuitive means of exploring relevant structures. This will allow suspicious groups to be examined by a domain expert, in order to confirm the existence of fraud and act upon it. The ability to guide the search to within a small subset of the customer base makes the potential detection of fraud more efficient and enables the presentation of results in a visually network-oriented and attribute-enriched way.

References

1. Garey, M.R., Johnson, D.S.: Computers and Intractability: A Guide to the Theory of NP-Completeness. W. H. Freeman & Co., New York (1979)
2. Subelj, L., Furlan, S., Bajec, M.: An expert system for detecting automobile insurance fraud using social network analysis. Expert Systems with Applications 38(1), 1039–1052 (2011)
3. Wan-Shiou Yang, S.Y.H.: A process-mining framework for the detection of healthcare fraud and abuse. Expert Syst. Appl., 56–68 (2006)
4. Ratkiewicz, J., Conover, M., Meiss, M., Gonçalves, B., Patil, S., Flammini, A., Menczer, F.: Truthy: mapping the spread of astroturf in microblog streams. In: Proceedings of the 20th International Conference Companion on World Wide Web (WWW 2011), pp. 249–252. ACM (2011)
5. Morselli, C., Giguère, C., Petit, K.: The efficiency/security trade-off in criminal networks. Social Networks 29(1), 143–153 (2007)
6. Reuter, P., Truman, E.M.: Chasing Dirty Money: The Fight Against Money Laundering. Peterson Institute (2004)
7. Cortes, C., Pregibon, D., Volinsky, C.: Communities of Interest. In: Hoffmann, F., Adams, N., Fisher, D., Guimarães, G., Hand, D.J. (eds.) IDA 2001. LNCS, vol. 2189, pp. 105–114. Springer, Heidelberg (2001)
8. Pandit, S., Chau, D.H., Wang, S., Faloutsos, C.: NetProbe: A Fast and Scalable System for Fraud Detection in Online Auction Networks. In: Proceedings of the 16th International Conference on World Wide Web (2007)
9. Fortunato, S.: Community Detection in Graphs. Physics Reports 486(3–5), 75–174 (2010)
10. Bron, C., Kerbosch, J.: Algorithm 457: Finding All Cliques of an Undirected Graph. Communications of the ACM 16(9), 575–577 (1971)
11. Seidman, S., Foster, B.: A Graph-Theoretic Generalization of the Clique Concept. Journal of Mathematical Sociology 6, 139–154 (1978)
12. Seidman, S.: Network Structure and Minimum Degree. Social Networks 5(3), 269–287 (1983)
13. Cohen, J.: Graph Twiddling in a MapReduce World. IEEE Computing in Science & Engineering 11(4), 29–41 (2009)
14. Cohen, J.: Trusses: Cohesive Subgraphs for Social Network Analysis. Technical report, National Security Agency (2008)
15. Holme, P., Saramäki, J.: Temporal networks. CoRR abs/1108.1780 (2011)
16. Vanetik, N., Gudes, E., Shimony, S.E.: Computing Frequent Graph Patterns from Semistructured Data. In: Proceedings of the International Conference on Data Minning, pp. 458–465 (2002)

17. Kuramochi, M., Karypis, G.: Grew: A Scalable Frequent Subgraph Discovery Algorithm. In: Proceedings of the IEEE International Conference on Data Mining (2004)
18. Kuramochi, M., Karypis, G.: Finding Frequent Patterns in a Large Sparse Graph. Data Mining and Knowledge Discovery 11, 243–271 (2005)
19. Yan, X., Yu, P., Han, J.: Graph Indexing: A Frequent Structure-Based Approach. In: Proceedings of the ACM SIGMOD International Conference on Management of Data, pp. 335–346 (2004)
20. Cheng, J., Ke, Y., Ng, W., Lu, A.: FG-Index: Towards Verification-Free Query Processing on Graph Databases. In: Proceedings of the ACM SIGMOD International Conference on Management of Data, pp. 857–872 (2007)
21. Xiong, L., Liu, L.: A reputation-based trust model for peer-to-peer ecommerce communities. In: Proceedings of the 4th ACM Conference on Electronic Commerce (EC 2003), pp. 228–229. ACM (2003)
22. Python Software Foundation: Python (2012), http://www.python.org
23. NetworkX Developers: NetworkX (2010), networkx.lanl.gov
24. Hunter, J., Dale D., Droettboom M.: matplotlib (2011), matplotlib.sourceforge.net
25. Gephi: Gephi (2012), gephi.org
26. Rivals, I., Personnaz, L., Taing, L., Potier, M.: Enrichment or depletion of a GO category within a class of genes: which test? Bioinformatics 23(4), 401–407 (2007)

Describing Locations Using Tags and Images: Explorative Pattern Mining in Social Media

Florian Lemmerich[1] and Martin Atzmueller[2]

[1] Artificial Intelligence and Applied Computer Science,
University of Würzburg,
97074 Würzburg, Germany
lemmerich@informatik.uni-wuerzburg.de
[2] Knowledge and Data Engineering Group,
University of Kassel
atzmueller@cs.uni-kassel.de

Abstract. This paper presents an approach for explorative pattern mining in social media based on tagging information and collaborative geo-reference annotations. We utilize pattern mining techniques for obtaining sets of tags that are specific for the specified point, landmark, or region of interest. Next, we show how these candidate patterns can be presented and visualized for interactive exploration using a combination of general pattern mining visualizations and views specialized on geo-referenced tagging data. We present a case study using publicly available data from the Flickr photo sharing platform.

1 Introduction

Given a specific location, it is often interesting to obtain representative and interesting descriptions for it, e.g., for planning touristic activities. In this paper, we present an approach for modeling location-based profiles of social image media by obtaining a set of relevant image descriptions (and their associated images) for a specific point of interest, landmark, or region, described by geo-coordinates provided by the user. We consider publicly available image data, e.g., from photo management and image sharing applications such as Flickr[1] or Picasa[2].

In our setting, each image is tagged by users with several freely chosen tags. Additionally, each picture is annotated with a geo-reference, that is, the latitude and the longitude on earth surface where the image was taken. Based on this information, we try to explore the collaborative tagging behavior in order to identify interesting and representative tags for a specific location of interest. This can be either a point or a region, so that the method can be applied both for macroscopic (regional) and microscopic (local) analysis. Furthermore, by appropriate tuning and a fuzzified focus, also mesoscopic analyses combining both microscopic and macroscopic views can be implemented.

---

[1] www.flickr.com
[2] www.picasa.com

M. Atzmueller et al. (Eds.): MSM/MUSE 2011, LNAI 7472, pp. 77–96, 2012.

Since the problem of identifying *interesting* and representative descriptions of a location is to a certain degree subjective, one cannot expect to identify the best patterns in a completely automatic approach. On the other hand, considering datasets with thousands of tags, manual browsing is usually not an option.

Therefore, we propose a two step approach for tackling this problem: The first step uses pattern mining techniques, e.g. [1, 2], to automatically generate a candidate set of potentially interesting descriptive tags. For this task, we present three different options for constructing target concepts. In the second step, a human explores this candidate set of patterns and introspects interesting patterns manually. In a user-guided environment, explorative pattern mining can then be applied iteratively adapting the process steps according to the analysis goals. Additionally, background knowledge regarding the set of tags can be easily incorporated in a semi-automatic process, such that new attributes are generated from tag hierarchies that can be manually refined and included in the process. To further improve the results, we propose a simple but effective method for incorporating a weighting schema to avoid bias towards very active users.

The presented approach is thus implemented in a semi-automatic way. In such contexts, typically advanced methods for the visualization and browsing of the respective tags sets are required according to the *Information Seeking Mantra* by Shneiderman [3]: *Overview, Zoom and Filter, Details on Demand.* We propose a set of techniques for exploring the statistics and spatial distribution of the candidate tags. These include visualizations adapted from statistics, from the area of pattern mining, and also domain specific views developed for spatial data. The presented approach is embedded into the comprehensive pattern mining and subgroup discovery environment VIKAMINE [4], which was extended with specialized plug-ins for handling and visualizing geo-spatial information.

From a scientific point of view, the tackled problem is interesting as it requires the combination of several distinct areas of research: Pattern mining, mining social media, mining (geo-)spatial data, visualization, knowledge acquisition and interactive data mining. Our contribution can be summarized as follows:

1. We adapt and extend pattern mining techniques to the mining of combined geo-information and tagging information.
2. To avoid bias towards users with very many resources, we propose a user weighting schema.
3. We show how background knowledge about similar tags can be included to define or refine topics consisting of multiple tags.
4. For the explorative mining approach we provide a set of visualizations.
5. The presented approach is demonstrated in a case study using publicly available data from Flickr with respect to two well-known locations in Germany.

The rest of the paper is structured as follows: Section 2 describes the candidate generation through pattern mining. After that, Section 3 introduces the interactive attribute construction and visualization techniques. Next, Section 4 features two real-world case studies using publicly available data from Flickr. Section 5 discusses related work. Finally, Section 6 concludes the paper with a summary and interesting directions for future research.

2 Location-Based Profile Generation and Interactive Exploration of Social Image Media

The problem of generating representative tags for a given set of images is an active research topic, see [5]. In contrast to previously proposed techniques, cf. [6], our approach does not require a separate clustering step. Furthermore, we also include interactive exploration into our overall discovery process: The approach starts by obtaining a candidate set of patterns from an automated pattern mining task. However, since it is difficult to extract exactly the most interesting patterns automatically, we propose an interactive and iterative approach: Candidate sets are presented to the user, who can refine the obtained patterns, visualize the patterns and their dependencies, add further knowledge, or adapt parameters for a refined search iteratively.

2.1 Background on Pattern Mining

Since the number of used tags in a large dataset usually is huge, it is rather useful to provide the user with a targeted set of interesting candidates for interactive exploration. For this task, we utilize the data mining method of pattern mining, specifically subgroup discovery [1, 2, 7, 8]. This allows us to identify not only interesting single tags efficiently, but also combinations of tags, which are used unusually more frequently together in a given area of interest.

Subgroup discovery aims at identifying interesting patterns with respect to a given target property of interest according to a specific interestingness measure. In our context, the target property is constructed using a user-provided location, i.e., a specific point of interest, landmark, or region, identified by geo-coordinates.

Pattern mining is thus applied for identifying relations between the (dependent) target concept and a set of explaining (independent) variables. In the proposed approach, these variables are given by (sets of) tags that are as specific as possible for the target location. The top patterns are then ranked according to the given interestingness measure.

Formally, a database $D = (I, A)$ is given by a set of individuals I (pictures) and a set of attributes A (i.e., tags). A *selector* or *basic pattern* $sel_{a=a_j}$ is a boolean function $I \rightarrow \{0, 1\}$ that is true, iff the value of attribute a is a_j for this individual. A (complex) *pattern* or *subgroup description* $sd = \{sel_1, \ldots, sel_d\}$ is then given by a set of basic patterns, which is interpreted as a conjunction, i.e., $sd(I) = sel_1 \wedge \ldots \wedge sel_d$. We call a pattern sd_s a generalization of its specialization sd_g, iff $sd_g \subset sd_s$. A subgroup (extension) sg is then given by the set of individuals $sg = ext(sd) := \{i \in I | sd(i) = true\}$ which are covered by the subgroup description sd.

A subgroup discovery task can now be specified by a 5-tuple (D, T, S, Q, k). The target concept $T : I \rightarrow \mathbb{R}$ specifies the property of interest. It is a function, that maps each instance in the dataset to a target value t. It can be binary (e.g., the instance/picture belongs to a neighborhood or not), but can use arbitrary target values (e.g, the distance of an instance to a certain point in space). The search space 2^S is defined by a set of basic patterns S. Given the dataset D

and target concept t, the quality function $Q: 2^S \to \mathbb{R}$ maps every pattern in the search space to a real number that reflects the interestingness of a pattern. Finally, the integer k gives the number of returned patterns of this task. Thus, the result of a subgroup discovery task is the set of k subgroup descriptions $res_1, \ldots, res_k$ with the highest interestingness according to the quality function. Each of these descriptions could be reformulated as a rule $res_i \to t$.

While a huge amount of quality functions has been proposed in literature, cf. [9], the most popular interesting measures trade-off the size $|ext(sd)|$ of a subgroup and the deviation $t - t_0$, where t is the average value of the target concept in the subgroup and t_0 the average value of the target in the general population. Please note, that for binary t the average value of t reflects the likelihood of t in the respective set. Thus, the most used quality functions are of the form

$$q_a(sd) = |ext(sd)|^a \cdot (t - t_0), a \in [0; 1]$$

For binary target concepts, this includes for example the *weighted relative accuracy* for the size parameter $a = 1$ or a simplified binomial function, for $a = 0.5$.

2.2 Target Concept Construction

The most critical issue for formulating the location-based tag mining problem as a pattern mining task is how to construct a proper target concept. In this paper we propose and discuss the effects of three different approaches: Using the *raw* distance, a parametrized neighborhood function, and a "fuzzified" neighborhood function.

First, we could use the raw distance of an image to the point of interest as a numeric target property. Given latitudes and longitudes the distance on the earth surface of any point $p = (lat_p, long_p)$ to the specified point of interest $c = (lat_c, long_c)$ can be computed by:

$$d(p) = r_e \cdot \arccos(\sin(lat_p) \cdot \sin(lat_c) + \cos(lat_p) \cdot \cos(lat_c) \cdot \cos(long_c - long_p)),$$

where r_e is the earth radius.

Using this as the numeric target concept, the task is to identify patterns, for which the average distance to the point of interest is very small. For example, the target concept for an interesting pattern could be described as: "Pictures with this tag are on average 25km from the specified point of interest, but the average distance for all pictures to the point of interest is 455 km".

The advantages of using the numeric target concept is that it is parameter-free and can be easily interpreted by humans. However, it is unable to find tags, which are specific to more than one location. For example, while for the location of the Berlin olympic stadium the tag "olympic" could be a regarded as specific. However, if considering other olympic stadiums (e.g., in Munich) the average distance for the tag "olympic" is quite large. Therefore, we define a second function: The neighborhood distance requires a maximum distance d_{max} to the location of interest. Then, the target concept is given by:

$$neighbor(p) = \begin{cases} 0, \text{ if } d(p) < dist_{max} \\ 1, \text{ else} \end{cases}$$

Tags are then considered as interesting, if they occur relatively more often in the neighborhood than in the total population. For example, the target concept for an interesting pattern in this case could be described as: "While only 1% of all pictures are in the neighborhood of the specified point of interest, 33% for pictures with tag x are in this neighborhood." The downside of this approach is however, that it is strongly dependent on the chosen parameter d_{max}. If this parameter is too large, then the pattern mining step will not return tags specific for the point of interest, but for the surrounding region. On the other hand, if d_{max} is too small, then the number of instances in the respective area is very low and thus can easily influenced by noise.

Therefore, the third considered approach is to "fuzzify" the second approach: Instead of a single distance d_{max} we define a minimum distance d_{lmax} and a maximum distance d_{umax} for our neighborhood. Images with a distance smaller than d_{lmax} are counted fully to the neighborhood but only partially for distances between d_{lmax} and d_{umax}. For the transition region between d_{lmax} and d_{umax} any strictly monotone function could be used. In this paper, we concentrate on the most simple variant, that is, a linear function. Alternatives could be sigmoid-functions like the generalized logistic curve.

$$fuzzy(p) = \begin{cases} 0, & \text{if } d(p) < d_{lmax} \\ \frac{d(p)-d_{lmax}}{d_{umax}-d_{lmax}}, & \text{if } d(p) > d_{lmax} \text{ and} \\ & d(p) < d_{umax} \\ 1, & \text{otherwise} \end{cases}$$

In doing so, we require one more parameter to chose, however, using such soft boundaries the results are less sensible to slight variations of the chosen parameters. Thus, we achieve a smooth transition between instances within or outside the chosen neighborhood.

Figure 1 depicts the described options: The fuzzy function can be regarded as a compromise between the other two function. It combines the steps for the neighborhood function with a linear part that reflects the common distance function.

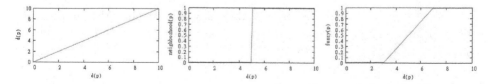

Fig. 1. The three proposed distance functions $d(p)$, $neighbor(p)$ with a threshold of $dist_{max} = 5$ and $fuzzy(p)$ with thresholds $d_- = 3$ and $d_+ = 7$ as a function over $d(p)$. It can be observed, that $d(p)$ is (obviously) linear, $neighbor(p)$ is a step function, and $fuzzy(p)$ combines both properties in different sections.

2.3   Avoiding User Bias: User–Resource Weighting

In the previously described process for candidate generation all images are treated as equally important. However, due to the common *power law* distribution between users and resources (images) in social media systems, only a few but very active users contribute a substantial part of the data. Since images from a specific user tend to be concentrated on certain locations and users also often apply a specific vocabulary, this can induce a bias towards the vocabulary of these active users. As an extreme example, consider a single "power user", who shared hundreds of pictures of a specific event at one location and tags all photos of this event with a unique name. Given the approach presented above this name is then considered as very important for that location, although the tag is not commonly used by the user base.

One possibility to solve this issue could be to utilize an interestingness measure that also incorporates the user count. That is, one could extend the standard quality function given above by adding a term, that reflects the number of different users that own a picture in the evaluated subgroup. Such an extended quality function could be defined as $q_a(sd) = |ext(sd)|^a \cdot (t - t_0) \cdot |u(sd)|$, where $|u(sd)|$ is the user count for images in the respective subgroup. Unfortunately, such interestingness measures are not supported by efficient exhaustive algorithms for subgroup discovery, e.g., SD-Map [10] or BSD [11]. On the other hand, more basic algorithms, for example exhaustive depth-first search without a specialized data structure scale not very well for the problem setting of this paper, with thousands of tags as descriptions and possibly millions of instances in an interactive setting.

Therefore, we propose to apply a slightly different approach to reduce user bias in our application. We assume that a single picture might be overall less important, if a user shared a large amount of images. This is implemented by applying an instance weight for each resource, that is, for each image in our application. Thus, when computing statistics of a subgroup the overall count and the target value, which is added if the respective image is part of a subgroup, is multiplied by the corresponding weight $w(i)$. The weight is smaller, if more pictures are contributed by the owner of the image. For our experiments we utilized a weighting function of

$$w(i) = \frac{1}{\sqrt{(|\{j|j \text{ is contributed by the user that contributed } i\}|)}} .$$

Instance weighting is supported by SD-Map as well as many other important subgroup discovery algorithms, since it is also applied in pattern set mining approaches such as weighted covering [7].

3   Interactive Exploration

In the following, we first describe the options for including background knowledge for semi-automatic attribute construction. After that, we describe the different visualization options.

3.1   Semi-automatic Attribute Construction

In social environments similar semantics are often expressed using diverse sets of tags, e.g., due to different languages. For an improved analysis it can be helpful to combine multiple tags into *topics* (meta-tags), that is, sets of semantically related attributes. The attribute hierarchy editor shown in Figure 2 allows an easy but fine-grained specification of topics by editing a text document using dash-trees [12] as a simple intuitive syntax: A tree structure can easily be defined by adding "-" characters at the start of the respective lines, see Figure 2. The root of the tree defines the topic name, the tree children declares included tags for this topic. For each topic a new attribute is constructed in the system, that is set to *true* for a single instance, iff at least one of the attributes identified by a child node is *true* in this instance. The hierarchies are directly specified in VIKAMINE and propagated to the applied dataset.

In addition to providing the knowledge purely manually, we can also apply a semi-automatic approach. This is implemented, e.g., using LDA-based approaches (*latent dirichlet allocation* [13]). LDA provides for a convenient data preprocessing option by capturing semantically similar tags and thus helps to inhibit the problem of synonyms, semantic hierarchies, etc. After that, the set of proposed topics

Fig. 2. Editor for specifying background knowledge (tag hierarchies) in textual form. The tag hierarchies can be generated, e.g., by LDA-based approaches, and can be refined in a semi-automatic step. In this example for instance the new attribute *cemetery** is constructed that is true, iff the respective image has been tagged by any of the tags beyond (*cemetery, friedhof, grave, cemeteries, cementerios, cimiteri, graves, friedhöfe, gräber*).

can then be tuned and refined by the user by utilizing dash-trees. In this way, we efficiently build interpretable tag clusters, i.e., for obtaining descriptive topic sets.

3.2 Visualization

In our approach, the problem of identifying tags specific for a region is formulated as a pattern mining task. While this task can generate candidate patterns, often only manual inspection by human experts can reveal the most informative patterns. This is especially the case, when considering that the interestingness is often subjective and dependent on prior knowledge.

As a simple example, if you knowingly choose a point of interest in the city of Berlin, the information, that the tag "berlin" is often used there, will not add much knowledge. However, if a point is chosen arbitrarily on the map without any information about the location, then the information that this tag is used frequently in that area is supposedly rather interesting. Therefore, we consider possibilities to interactively explore, analyze and visualize the candidate tags and tag combinations as essential for effective knowledge discovery in our setting. We consider three kinds of visualizations:

1. Traditional visualizations from exploratory data analysis are mainly used for introspection of candidate patterns. Typical visualizations include the contingency table, pie charts, and box plots. An especially important visualization of this category proved to be a distance histogram. This histogram shows on the x-axis the distances $d(p)$ from the location of interest and on the y-axis the number of images with the specified tag(s) at that distance.
2. For an interactive exploration of the mined profiles and the tag sets and for comparative visualization we can utilize various established techniques for interactive subgroup mining, cf. [4]. These user interfaces include for example:
   (a) The *Zoomtable* which is used to browse over on the refinements of the currently selected pattern. For numeric targets, it includes the distribution of tags concerning the currently active pattern. For the binary 'neighbor' target concept, it shows more details within the zoom bars, cf. [4], e.g., showing the most interesting factors (tags) for the current pattern and target concept.
   (b) The *nt-Plot* compares the size and target concept characteristics of many different pattern. In this ROC-space related plot, e.g., [4], each pattern is represented by a single point in two dimensional space. The position on the x-axis denotes the size of the subgroup, that is, the number of pictures covered by the respective tags. The position on the y-axis describes the value of the target concept for the respective pattern.

   Thus, a pattern with a high frequency that is not specific for the target location is displayed on the lower right corner of the plot, while a very specific tag, which was not frequently used is displayed on the upper left corner.

(c) The *Specialization Graph* is used to show the dependencies between tag combinations, cf. [14]. In this graph, each pattern is visualized by a node in the graph. Each node is represented by a two-part bar. The total length of these bars represents the number of cases covered by this pattern, while the ratio between the two parts of the bar represent the value/share of the target concept within the extension of the pattern. Generalization relations between patterns are depicted by directed edges from more general to more specific patterns. For example, the patterns *arts* and *arts ∧ night* are connected by an edge pointing at the latter patterns.

For a more specific exploration of the location-based profiles of social image media specialized visualization methods can furthermore be exploited:

(a) The *Distance Attribute Map* is a view, that allows for the interactive creation of distance attributes ($d(p)$, $neighbor(p)$ and $fuzzy(p)$) by selecting a point p on a dragable and zoomable map. Future improvements could incorporate online search function, e.g., by using the Google Places API.

(b) The *Tag Map* visualizes the spatial distribution of tags on a dragable and zoomable map. Each picture for a specific pattern is represented by a marker on the map. Since for one pattern easily several thousand pictures could apply, we recommend to limit the number of displayed markers. In our case study (see Section 4) we chose a sample of at most 1000 markers. In a variant of this visualization also the distribution of sets of tags can be displayed on a single map in order to compare their distributions. An exemplary zoomed-in Tag-Map for the tags *brandenburgertor* and *holocaust* (for the memorial) is shown in Figure 3.

(c) The *Exemplification View* displays sample images for the currently displayed tag. This is especially important, since pattern exemplification has shown to be essential for many applications, e.g., [15]. Using this view, the overall application can be used to not only browse and explore the used tags with respect to their geo-spatial distribution, but also

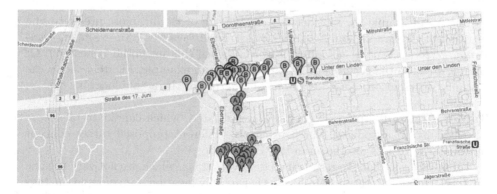

Fig. 3. Example comparative Tag-Map visualization from the case study (zoomed in): Pictures with tag "holocaust" are marked with an red "A", while pictures for the tag "brandenburgertor" are marked with a green "B".

allows for interactive browsing of the images itself. Since there are possibly too many pictures described a set of tags to be displayed at once, we propose to select the shown images also with respect to their popularity, i.e., the number of views of the images, if this information is available.

The interactive exploration also can utilize background knowledge concerning the provided tags, which is entered either in a textual or graphical form.

The proposed features were implemented as a plugin for the interactive subgroup discovery environment VIKAMINE[3]. For incorporating the traditional plots the VIKAMINE R-Plugin was used as a bridge to the R[4] language for statistical computing.

4   Case Study: Flickr

We show the effectiveness of our approach in two case studies. These application scenarios utilize 1.1 million images collected from Flickr. We selected those that were taken in 2010 and are geotagged with a location in Germany.

For the collected tagging data, we applied data cleaning and preprocessing methods, e.g., stemming. We considered all tags that were used at least 100 times. This resulted in about 11,000 tags. In the case studies we show how the combination of automated pattern mining, visualization and specialized views for geo-referenced tagging data enables the identification of tag combinations which are interesting for the specified location. For pattern mining, we applied the proposed quality function with $a = 0.5$.

For our case studies, we present results for two example locations: The famous Brandenburger Tor in Berlin and the Hamburg harbor area. The goal was to enable the identification of tags, which are representative especially for this region, for people without knowledge of the respective location.

4.1   Example 1: Berlin, Brandenburger Tor

In our first example we consider the city centre of Berlin, more precisely, the location of the Brandenburger Tor. The expected tags were, for example, *brandenburgertor*, *reichstag*, *holocaustmemorial* (since this memorial is nearby). Of course, also the tag *berlin* is to be expected. As an example, Figure 4 shows the distance distribution of the tag *brandenburgertor* to the actual location.

Target Concept Options. First we investigated, which candidate tags were returned by an automatic search using the different proposed target concept options. The results are shown in the Tables 1-5.

Table 1 shows, that the results include several tags, which are not very specific for the location of interest, but for another nearby location, for example the tags *Potsdam* or *Leipzig* for cities close to Berlin. This can be explained by the fact,

---

[3] www.vikamine.org

[4] www.r-project.org

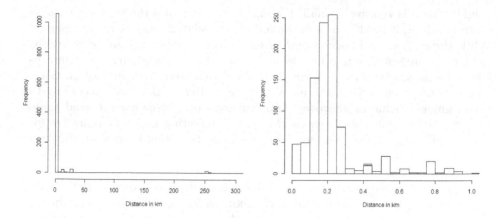

Fig. 4. Histogram showing the distances of pictures with the tag "brandenburgertor" to the actual location. It can be seen in the left histogram that the tag is very specific, since the vast majority of pictures with this tag is within a 5km range of the location. The histogram on the right side shows the distance distribution up to 1km in detail. It can be observed that most pictures are taken at a distance of about 200m to the sight.

Table 1. Brandenburger Tor: top patterns (max. description size 1) for the common mean distance target function

Tag	Subgroup Size	Mean Target Distance (km)
berlin	113977	10.48
potsdam	5533	26.83
brandenburg	5911	47.33
charlottenburg	4738	10.90
art	24067	211.28
leipzig	10794	147.87
kreuzberg	3935	14.11
nachbarn	3691	6.16
leute	4547	53.37
strassen	6899	126.83
berlinmitte	3054	4.76

that these tags are quite popular and the average distance for pictures with this tag is relatively low in comparison to the total population even if pictures do not correspond to the location of interest itself, but for a nearby location. Since the use of the distance function $d(p)$ does not allow for parametrization, it is difficult to adapt the search, such that those tags are excluded.

Tables 2-4 show the *neighbor* function with different distance thresholds d_{max}, from 0.1 km to 5 km. The results for this target concept are strongly dependent on this threshold. For a very small value of $d_{max} = 0.1\ km$ the results seem to be strongly influenced by some kind of noise, since the number of pictures in that neighborhood is relatively small. For example it includes the tags *metro*, *gleis* (translated: "rail track") or *verkehrsmittel* (translated "means of transport"). While these tags should occur more often in urban areas, they are by no means the most representative tags for the area around the Brandenburger Tor. In contrast, the parameter $d_{max} = 1\ km$ yields results that do meet our expectations. The resulting tags reflects the most important sites in that area according to travel guides, including *reichstag*, *brandenburgertor*, *potsdamerplatz* and *sonycenter*. We consider these tags as the most interesting and representative for this given location. However, we do not assume that this parameter will lead to the best result in all circumstances. For example, in more rural areas, where more landscape pictures with a larger distances to depicted objects are taken, we expect that a larger value of d_{max} might be needed. As shown in Table 4, for a parameter of $d_{max} = 5\ km$ the results show to be tags, which are specific for Berlin as a whole, but not necessarily for the area around the Brandenburger Tor. The results include tags like *tiergarten*, *kreuzberg* or *alexanderplatz* which describe other areas in Berlin.

Finally, Table 5 shows the fuzzified distance function, ranging from 1km to 5km as lower and upper thresholds. The results indicate, that this function is less sensitive to the parameter choices. Therefore, selecting the parameter is less difficult since, e.g., distances like 1-5km as in the presented example can be applied for a microscopic to a mesoscopic perspective. The collected results form a nice compromise between the results of the *neighbor* functions.

Including Instance Weighting. Taking a closer look at the results of Table 4 most of the resulting tags provide a good description of the larger area of Berlin. However, there are a few exceptions: *karnevalderkulturen* describes a seasonal well known, but not indicative event in Berlin. *heinrichböllstiftung* is a political foundation, for which the headquarters are located in Berlin. While both tags are certainly associated with Berlin, one would not expect them to be as important or typical for Berlin as other descriptions. The occurrence of these tags can be explained by a few "power users" that extensively used these tags for many images. To show this effect, we added an additional column for to Table 4, which notes the overall count of users that used that description. For example the tag *heinrichböllstiftung* was applied for 1211 images, but only by three different users. To avoid such results in the candidate generation, we apply an instance (resource) weighting as described in Section 2.3. The results are presented in Table 6. The table shows, for example that the tags *heinrichböllstiftung* and *karnevalderkulturen* have disappeared and are replaced by more broadly used descriptions of Berlin attractions such as *fernsehturm* (translated: television tower) or *memorial* (for the previously mentioned holocaust memorial). Thus, we consider the attribute weighting as appropriate to reduce bias towards the vocabulary of only a few but very active users, as shown in the example.

Attribute Construction. As can been seen from this example (Table 4), the automatic candidate generation tends to return semantically equivalent or very closely related tags in the results, i.e. translations of tags into other languages, for example *berlin, berlino and berlijn*. Such results fill slots in the result set of the candidate generation, suppress further interesting and make the results more difficult to comprehend. Additionally, one wants to perform the next step of the analysis— the interactive exploration — for these descriptions at once. In order to identify such equivalent tags and combine them within the system we used our semi-automatic attribute construction technique. To do so, first a latent dirichlet allocation is performed on the dataset to obtain a set of 100 candidate topics. The results were manually evaluated and transformed in a dash-tree format, see Section 3.1. The input format was then used to construct new combined tags (topics) that are treated like regular tags. Additionally, the tags that were used to build these meta-tags were excluded from candidate generation

The automatically constructed tags were of mixed quality: For a few topics the describing tags could be almost directly used as equivalent tags. For example, one resulting topic of the LDA was given by the tags: *cemetery, friedhof, grave, cimetičre, cemeteries, cementerio, friedhöfe, cementerios, cemitério, cimiteri, cimetičres, cemitérios* and *graves*. The majority of the topics included several tags that can be considered as equivalent, but include other tags as well, for example: *architecture, building, architektur, church, dom, cathedral, germany, tower, gebäude, window, glass*. Some of these tags can be used to construct a new meta-tag by manual refinement, e.g. *architecture, building* and *architektur*, however the tags *germany* or *glass* should not be used for this purpose. The last group of topics consisted of rather loosely related tags, for example: *winter, thuringia, snow, town, tree, village, sky*. These topics were considered inappropriate for the purpose of constructing expressive attributes.

In summary, LDA provided for a very good starting point to find equivalent tags. However, applying only the automatic method was far from a quality level that enabled us to use the results directly to construct clear meaningful and comprehensible combined tags. The text-based format in our mining environment

Table 2. Brandenburger Tor: top patterns (description size 1) for the target concept function *neighbor*, with $d_{max} = 0.1$ km

Tag	Subgroup Size	Target Share
wachsfigur	322	0.99
madametussauds	177	0.853
celebrity	345	0.435
verkehrsmittel	163	0.313
metro	469	0.277
berlinunderground	158	0.247
kitty	185	0.227
brandenburgertor	1136	0.085
u55	114	0.263
ubahn	4295	0.034
unterdenlinden	573	0.075
gleis	375	0.085
bahnsteig	551	0.058

Table 3. Brandenburger Tor: top patterns (description size 1) for the target concept function *neighbor*, with $d_{max} = 1$ km

Tag	Subgroup Size	Target Share
berlin	113977	0.225
reichstag	2604	0.829
potsdamerplatz	2017	0.797
heinrichböllstiftung	1211	0.988
berlino	4162	0.461
brandenburgertor	1136	0.816
sonycenter	803	0.923
gendarmenmarkt	696	0.885
potsdamer	577	0.88
bundestag	1096	0.611
brandenburggate	643	0.776
brandenburger	401	0.913
friedrichstrasse	558	0.735
unterdenlinden	573	0.705
panoramapunkt	271	1
holocaustmemorial	301	0.93

Table 4. Brandenburger Tor: top patterns (description size 1) for the target concept function *neighbor* and a threshold $d_{max} = 5$ km. The last column shows the overall count of users that used this description.

Tag	Subgroup Size	Target Share	Users
berlin	113977	0.745	5703
kreuzberg	3933	0.961	405
berlino	4162	0.915	392
mitte	3507	0.972	404
reichstag	2604	0.976	680
berlinmitte	3053	0.832	96
potsdamerplatz	2017	0.97	375
hauptstadt	2350	0.892	106
karnevalderkulturen	1851	0.958	36
alexanderplatz	1699	0.989	546
berlijn	2094	0,844	120
berlinwall	1635	0.914	275
graffiti	6136	0.525	838
tiergarten	2497	0.749	287
berlín	1431	0.931	119
heinrichböllstiftung	1211	1	3

proved to be easy to use and well-fit for this purpose. The automatic method (LDA) proposed suitable sets of tags which could be manually refined. Depending on the amount of total tags this requires a certain amount of manual work. Accordingly, the decision, which tags can be considered semantically equivalent is also subjective to a certain degree. Nonetheless, this only emphasizes the need

Table 5. Brandenburger Tor: top patterns (description size 1) for the 'fuzzified' target concept distance function ranging from 1 km to 5 km

Tag	Subgroup Size	Mean Target Share
berlin	113977	0.46
reichstag	2604	0.05
potsdamerplatz	2017	0.05
mitte	3507	0.42
berlinmitte	3053	0.30
heinrichböllstiftung	1211	0.01
hauptstadt	2350	0.34
brandenburgertor	1136	0.10
alexanderplatz	1699	0.28
city	18246	0.76
tiergarten	2497	0.42
platz	2171	0.4
touristen	2815	0.47
nachbarn	3691	0.55
sonycenter	803	0.02

Table 6. Brandenburger Tor: top patterns (description size 1) using instance weighting for the target concept function *neighbor* and a threshold $d_{max} = 5$ km. The last column shows the overall count of users that used this description

Tag	Subgroup Size	Target Share	Users
berlin	13790.6	0,804	5703
berlino	806.2	0,916	392
reichstag	431.9	0,972	680
mitte	366.3	0,97	404
kreuzberg	371	0,96	405
alexanderplatz	275.6	0,982	546
berlinwall	237.8	0,945	275
berlijn	291.7	0,85	120
fernsehturm	310.8	0,794	725
berlín	224.9	0,908	119
potsdamerplatz	196.4	0,963	375
wall	548.6	0,597	959
memorial	287.7	0,721	488
eastsidegallery	155.6	0,922	156
graffiti	661.6	0,506	838
brandenburgertor	139.4	0,931	332

of a simple interactive environment that enables also system users without a data mining background to combine attributes as they see them fit. This technique of attribute construction also enables the user to investigate self-constructed topics by interactive exploration by just creating a meta tag with certain selected keywords.

4.2   Example 2: Hamburg Harbor - "Landungsbrücken"

The second example considers the Hamburg harbor, especially the famous "Landungsbrücken". For this location, Figure 6 shows the distribution of several interesting tags in the zoomtable.

For the Hamburg example, we also show complex patterns, i.e., combinations of tags, in the result tables. Table 7 shows the results of applying the standard mean distance target concept, while Table 8 shows the results of the fuzzified target concept, ranging from 1km to 5km (lower, upper parameters).

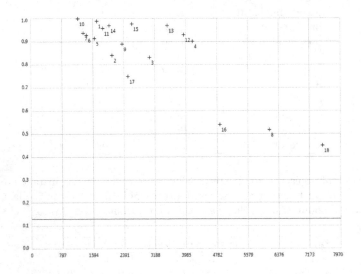

Fig. 5. An exemplary nt-plot for the location Brandenburgertor, for tags with a maximum distance of 5km. Tags that were used more often are shown on the right side of the diagramm, for example, "streetart" (16), "graffiti" (8), or "urban" (18). Tags that are very specific for the given target concept, that is, within a 5km area of the Berlin Brandenburger Tor, are displayed at the top of the diagramm. For example, the tag "urban" (18) was used relatively often, but it is not specific for the specified location of interest. However, tags such as "heinrichböllstiftung" (10), "alexanderplatz" (1), or "potsdamerplatz" (14) are very specific (and interesting) for the specified location.

Attributes	Values	
elbe	t	
fluss	f	t
hafen	f	t
hafencity	f	t
hamburg	t	
hansestadt	f	

Zoomtable ⊠ Subgroup Workspace

Fig. 6. The zoomtable showing some tags from the Hamburg Harbor

Table 7. Hamburg Harbor: The top patterns (max. description size 2) for the mean distance target concept

Tag	Subgroup Size	Mean Target Distance (km)
hamburg	29448	9.60
niedersachsen	34672	170.05
berlin	116979	258.34
schleswigholstein	9068	96.75
2010 AND hamburg	5255	7.81
oldenburg	10000	120.02
berlin AND germany	43280	256.95
ostsee	9565	154.41
hannover	8052	138.62
bremen	5656	99.06
lingen	14004	210.85
lingen AND germany	13909	210.82

Table 8. Hamburg Harbor: The top patterns (max. description size 2) for the 'fuzzified' target concept distance function ranging from 1 km to 5 km.

Tag	Subgroup Size	Mean Target Share
hamburg	29448	0,89
deutschland AND hamburg	6127	0.80
hafen AND hamburg	2163	0.69
hansestadt AND hamburg	1376	0.60
deutschland AND hansestadt	1676	0.68
elbe AND hamburg	1786	0.70
schiff AND hamburg	996	0.58
hafen AND elbe	656	0.52
hansestadt	2906	0.81
ship AND hamburg	882	0.63

It is easy to see, that these results support the findings for the Berlin example: The fuzzified approach is more robust and concentrates on the important tags well, while the standard approach is suitable on a very macroscopic scale. It includes tags that are specific for the region, e.g., *schleswigholstein* or relatively close cities such as *Lingen* and *Hannover*.

5   Related Work

This paper combines approaches from three distinct research areas, that is, pattern mining, mining (geo-)spatial data, and mining social media. First, in contrast to the common pattern mining approaches, we introduce different target concept (functions), extending the traditional definition of target concepts.

Next, (geo-)spatial data mining [16] aims to extract new knowledge from spatial databases. In this context, often established problem statements and methods have been transfered to the geo-spatial setting, for example, considering association rules [17]. In this paper, we incorporate geo-spatial elements to construct distance-based target concepts according to different intuitions. Also, for the combination of pattern mining and geo-spatial data, we provide a set of visualizations and interactive browsing options for a semi-automatic mining approach.

Regarding mining social media, specifically social image data, there have been several approaches, and the problem of generating representative tags for a given set of images is an active research topic, see e.g. [5]. Sigurbjörnsson and van Zwol also analyze Flickr data and provide a characterization of how users apply tags and which information is contained in the tag assignments [18]. Their approach is embedded into a recommendation method for photo tagging, similar to [19] who analyze different aspects and contexts of the tag and image data. Abbasi et al. present a method to identify landmark photos using tags and social Flickr groups [20]. They apply group information and statistical preprocessing of the tags for obtaining interesting landmark photos.

In contrast to previously proposed techniques, e.g., [6], our approach does not require a separate clustering step. Furthermore, we focus on descriptive patterns consisting of tags that are interesting for a specific location; the interestingness can also be flexibly scaled by tuning the applied quality function. In contrast to the above automatic approaches, we also present and extend different visualizations for a semi-automatic interactive approach, integrating the user.

6   Conclusions

In this paper, we have presented an approach for obtaining location-based profiles for social image media using explorative pattern mining techniques. Candidate sets of tags, which are specific for the target location are mined automatically by an adapted pattern mining search step and can be refined subsequently. The approach enables several options including selectable analysis-specific interestingness measures and semi-automatic feature construction techniques. In an interactive process, the results can then be visualized, introspected and refined. For demonstrating the applicability and effectiveness, we presented a case study using real-world data from the photo sharing application Flickr considering two well-known locations in Germany.

For future work, we aim to consider richer location descriptions as well as further descriptive data besides tags, e.g., social friendship links in the photo sharing application, or other link data from social networks. Also, the integration of information extraction techniques, see for example [21], seems promising, in order to add information from the textual descriptions of the images. Furthermore, we plan to include more semantics concerning the tags, such that a greater detail of relations between the tags can be implemented in the preprocessing, the mining, and the presentation.

Acknowledgment. This work has partially been supported by the VENUS research cluster at the interdisciplinary Research Center for Information System Design (ITeG) at Kassel University, and by the EU project EveryAware.

References

1. Wrobel, S.: An Algorithm for Multi-Relational Discovery of Subgroups. In: Komorowski, J., Żytkow, J.M. (eds.) PKDD 1997. LNCS, vol. 1263, pp. 78–87. Springer, Heidelberg (1997)
2. Atzmueller, M., Lemmerich, F.: Fast Subgroup Discovery for Continuous Target Concepts. In: Rauch, J., Raś, Z.W., Berka, P., Elomaa, T. (eds.) ISMIS 2009. LNCS, vol. 5722, pp. 35–44. Springer, Heidelberg (2009)
3. Shneiderman, B.: The Eyes Have It: A Task by Data Type Taxonomy for Information Visualizations. In: Proc. IEEE Symposium on Visual Languages, Boulder, Colorado, pp. 336–343 (1996)
4. Atzmueller, M., Puppe, F.: Semi-Automatic Visual Subgroup Mining using VIKAMINE. Journal of Universal Computer Science (JUCS), Special Issue on Visual Data Mining 11(11), 1752–1765 (2005)
5. Liu, Z.: A Survey on Social Image Mining. In: Chen, R. (ed.) ICICIS 2011 Part I. CCIS, vol. 134, pp. 662–667. Springer, Heidelberg (2011)
6. Kennedy, L., Naaman, M.: Generating Diverse and Representative Image Search Results for Landmarks. In: Proceeding of the 17th International Conference on World Wide Web, pp. 297–306. ACM (2008)
7. Lavrac, N., Kavsek, B., Flach, P., Todorovski, L.: Subgroup Discovery with CN2-SD. Journal of Machine Learning Research 5, 153–188 (2004)
8. Atzmueller, M., Puppe, F., Buscher, H.P.: Exploiting Background Knowledge for Knowledge-Intensive Subgroup Discovery. In: Proc. 19th Intl. Joint Conf. on Artificial Intelligence (IJCAI 2005), Edinburgh, Scotland, pp. 647–652 (2005)
9. Geng, L., Hamilton, H.J.: Interestingness Measures for Data Mining: A Survey. ACM Computing Surveys 38(3) (2006)
10. Atzmüller, M., Puppe, F.: SD-Map – A Fast Algorithm for Exhaustive Subgroup Discovery. In: Fürnkranz, J., Scheffer, T., Spiliopoulou, M. (eds.) PKDD 2006. LNCS (LNAI), vol. 4213, pp. 6–17. Springer, Heidelberg (2006)
11. Lemmerich, F., Rohlfs, M., Atzmueller, M.: Fast discovery of relevant subgroup patterns. In: Proc. 23rd FLAIRS Conference (2010)
12. Reutelshoefer, J., Baumeister, J., Puppe, F.: Towards Meta-Engineering for Semantic Wikis. In: 5th Workshop on Semantic Wikis: Linking Data and People, SemWiki 2010 (2010)
13. Blei, D.M., Ng, A.Y., Jordan, M.I.: Latent Dirichlet Allocation. Journal of Machine Learning Research 3, 993–1022 (2003)
14. Klösgen, W., Lauer, S.R.W.: 20.1: Visualization of Data Mining Results. In: Handbook of Data Mining and Knowledge Discovery. Oxford University Press, New York (2002)
15. Atzmueller, M., Puppe, F.: A Case-Based Approach for Characterization and Analysis of Subgroup Patterns. Journal of Applied Intelligence 28(3), 210–221 (2008)
16. Koperski, K., Han, J., Adhikary, J.: Mining Knowledge in Geographical Data. Communications of the ACM 26 (1998)
17. Appice, A., Ceci, M., Lanza, A., Lisi, F., Malerba, D.: Discovery of Spatial Association Rules in Geo-Referenced Census Data: A Relational Mining Approach. Intelligent Data Analysis 7(6), 541–566 (2003)

18. Sigurbjörnsson, B., van Zwol, R.: Flickr Tag Recommendation based on Collective Knowledge. In: Proceeding of the 17th International Conference on World Wide Web, WWW 2008, pp. 327–336. ACM, New York (2008)
19. Lindstaedt, S., Pammer, V., Mörzinger, R., Kern, R., Mülner, H., Wagner, C.: Recommending Tags for Pictures Based on Text, Visual Content and User Context. In: Proc. 3rd International Conference on Internet and Web Applications and Services, pp. 506–511. IEEE Computer Society, Washington, DC (2008)
20. Abbasi, R., Chernov, S., Nejdl, W., Paiu, R., Staab, S.: Exploiting Flickr Tags and Groups for Finding Landmark Photos. In: Boughanem, M., Berrut, C., Mothe, J., Soule-Dupuy, C. (eds.) ECIR 2009. LNCS, vol. 5478, pp. 654–661. Springer, Heidelberg (2009)
21. Atzmueller, M., Beer, S., Puppe, F.: Data Mining, Validation and Collaborative Knowledge Capture. In: Brüggemann, S., d' Amato, C. (eds.) Collaboration and the Semantic Web: Social Networks, Knowledge Networks and Knowledge Resources. IGI Global (2011)

Learning and Transferring Geographically Weighted Regression Trees across Time

Annalisa Appice, Michelangelo Ceci, Donato Malerba, and Antonietta Lanza

Dipartimento di Informatica, Università degli Studi di Bari Aldo Moro,
via Orabona, 4 - 70126 Bari, Italy
{appice,ceci,malerba,lanza}@di.uniba.it

Abstract. The Geographically Weighted Regression (GWR) is a method of spatial statistical analysis which allows the exploration of geographical differences in the linear effect of one or more predictor variables upon a response variable. The parameters of this linear regression model are locally determined for every point of the space by processing a sample of distance decay weighted neighboring observations. While this use of locally linear regression has proved appealing in the area of spatial econometrics, it also presents some limitations. First, the form of the GWR regression surface is globally defined over the whole sample space, although the parameters of the surface are locally estimated for every space point. Second, the GWR estimation is founded on the assumption that all predictor variables are equally relevant in the regression surface, without dealing with spatially localized collinearity problems. Third, time dependence among observations taken at consecutive time points is not considered as information-bearing for future predictions. In this paper, a tree-structured approach is adapted to recover the functional form of a GWR model only at the local level. A stepwise approach is employed to determine the local form of each GWR model by selecting only the most promising predictors. Parameters of these predictors are estimated at every point of the local area. Finally, a time-space transfer technique is tailored to capitalize on the time dimension of GWR trees learned in the past and to adapt them towards the present. Experiments confirm that the tree-based construction of GWR models improves both the local estimation of parameters of GWR and the global estimation of parameters performed by classical model trees. Furthermore, the effectiveness of the time-space transfer technique is investigated.

1 Introduction

A main assumption underpinning geographic thinking is spatial non-stationarity, according to which a phenomenon varies across a landscape. In a regression task, where the predictor variables and the response variable are collected at several locations across the landscape, the major consequence of spatial non-stationarity is that the relationship between the predictor variables and the response variable is location-dependent. The consequence of this spatial variability is that a spatial analyst is discouraged from employing any conventional regression-based

M. Atzmueller et al. (Eds.): MSM/MUSE 2011, LNAI 7472, pp. 97–117, 2012.

model which assumes the independence of observations from the spatial location. Indeed, LeSage and Pace [19] have shown that the application of conventional regression models leads to wrong conclusions in spatial analysis and generates spatially autocorrelated residuals. One of the best known approaches to spatial regression is GWR (Geographical Weighted Regression) [2], a spatial statistics technique which addresses some challenges posed by spatial non-stationarity. In particular, GWR maps a local model as opposed to the global linear model conventionally defined in statistics, in order to fit the relationship between predictor variables and a response variable. In fact, unlike the conventional regression equation which defines single parameter estimates, GWR generates a *parametric* linear equation, where parameter estimates vary from location to location across the landscape. Each set of parameters is estimated on the basis of distance-weighted neighboring observations. The choice of a neighborhood is influenced by the observation that the positive spatial autocorrelation of a variable is common to many geographical applications [14]. In particular, the positive spatial autocorrelation of the response variable occurs when the response values taken at pairs of locations a certain distance apart (neighborhood) are be more similar than expected for randomly associated pairs of observations [18].

The focus of our attention on GWR is motivated by a number of recent publications which have demonstrated that this local spatial model is appealing in areas of spatial econometrics, including climatology [6], social segregation [20], industrialisation [15] as well as environmental and urban planning [28]. Despite of this, there are still research issues which are not faced by GWR. In the following we introduce a novel local algorithm which aims to solve them.

The main issue of GWR is that it outputs a single parametric equation, which represents the linear combination of all the predictor variables in the task, and considers the coefficients of the linear combination as parameters for the local estimation. This means that GWR assumes that predictor variables are all equally relevant for the response everywhere across the landscape, although it admits spatially varying parameters. Consequently, GWR does not deal with the spatially localized phenomenon of *collinearity*. In general, collinearity is a statistical phenomenon in which two or more predictor variables in a multiple regression model are highly linearly correlated. In this case, the coefficient estimates may change erratically in response to small changes in the model or the data, thus decreasing the predictive accuracy. In conventional regression, the problem of collinearity is addressed by identifying the subset of the relevant predictor variables and outputting the linear combination of only the variables in this subset [11]. Based on this idea, we argue that a solution to the spatial collinearity in GWR is to determine a parametric regression surface, which linearly combines a subset of the predictor variables. As we expect that variables in the subset may vary in space, we define a new spatially local regression algorithm, called GWRT (*G*eographically *W*eighted *R*egression *T*rees learner), which integrates a spatially local regression model learner with a tree-based learner. The tree-based learner recursively segments the landscape along the spatial dimensions (e.g. latitude and longitude), according to a measure of the positive spatial autocorrelation over the response values. In practice, the leaves of an induced tree represent a segmentation of the landscape into non-overlapping areal units which spatially reference response values positively autocorrelated within the corresponding leaf. The high concentration of autocorrelated response values falling in a leaf motivates the search for a parametric surface equation to be associated to the leaf. The leaf surface reasonably combines only a subset of relevant predictor variables and the parameters of this surface are locally estimated across the leaf. In particular, at each leaf, the predictor variables and the local parameters are learned by adapting the forward stepwise approach [11] defined for a global aspatial models to local spatial model learning.

Another important issue of both GWR and GWRT is that they do not capitalize on the *time dependence* among observations repeatedly collected across the same landscape at consecutive time points. This issue cannot be neglected due to the ubiquity of sensor network applications which continuously feed an unbounded amount of georeferenced and timestamped data (for instance, the temperature is periodically measured by weather stations across the Earth's surface). We face this issue by tailoring a transfer learning technique which adapts predictions obtained by geographically weighted regression trees learned by GWRT in the recent past towards the present. The research problem we consider focuses on applying knowledge from one set of past instances of a task to improve the performance of learning the same task in the present [26]. In our case the transferred model will reflect the spatial non-stationarity of phenomenon at present and also the time dependence among consecutive observations of the same phenomenon. For the transfer process, we sample few training key data in the present which are regularly distributed across the landscape. For each key observation, a transfer observation is computed with one predictor variable for each GWRT tree to be transferred from the past. The transferred model is a (spatially piecewise) regression model learned from these transfer data.

Therefore, the innovative contributions of this work with respect to original formulation of GWR are highlighted as follows. We propose a tree-based learner which allows the segmentation of the landscape in non-overlapping areal units that group positively autocorrelated response values. We do not assume any global form of the geographically weighted regression model, but we allow the variation across the landscape of the subset of predictive variables included the model. We design a stepwise technique to determine a geographically weighted regression model, where only the most promising predictive variables are selected. We define a transfer technique which allows us to use geographically weighted regression trees previously learned on past source domain data in order to improve the accuracy of prediction over the target domain data. We empirically prove that geographically weighted regression trees allow us a more accurate prediction of unknown response values spread across the landscape than three competitive methods: the traditional spatial statistic predictor GWR, the inductive aspatial model tree learner M5' [30] and the transductive spatial regression learner SpReCo [4] . Finally, we evaluate the viability of the transfer technique in a real application.

The paper is organized as follows. In the next Section we revise related work on regression in spatial statistics and spatial data mining as well as related work on inductive transfer learning. In Section 3, we illustrate the problem setting and introduce some preliminary concepts. In Section 4, we present the geographically weighted regression tree induction algorithm. In Section 5, we present the inductive transfer of this kind of tree-based models learned in the recent past to the present. In Section 6 we describe experiments we have performed with several benchmark spatial data collections. Finally, we draw some conclusions and outline some future work.

2   Background and Related Work

In order to clarify the background of this work, in this Section we illustrate related research on regression in both spatial data analysis and transfer learning.

2.1   Spatial Regression

Several definitions of the regression task have been formulated in spatial data analysis over the years. The formulation we consider in this work is the traditional one, where a set of attribute-value observations for the predictor variables and the response variable are referenced at point locations across the landscape. So far, several techniques have been defined to perform this task, both in spatial statistics and spatial data mining. A brief survey of these techniques (e.g. k-NN, geographically weighted regression, kriging) is reported in [27].

In particular, the k-Nearest Neighbor (k-NN) algorithm [23] is a machine learning technique which appears to be a natural choice for dealing with the regression task in spatial domains. Each test observation is labeled with the (weighted) mean of the response values of neighboring observations in the training set. A distance measure is computed to determine neighbors. As spatial coordinates can be used to determine the Euclidean distance between two positions, k-NN predicts the response value at one position by taking into account the observations which fall in the neighborhood. Thus, k-NN takes into account a form of positive autocorrelation over the response attribute only.

GWR [2] is a spatial statistic technique which extends the regression framework defined in conventional statistics by rewriting a globally defined model as a locally estimated model. The global regression model is a linear combination of predictor variables, defined as: $y = \alpha + \sum_{k=1}^{n} \beta_k x_k + \epsilon$ with intercept α and parameters β_k globally estimated for the entire landscape, by means of the least square regression method [11]. Then GWR rewrites this equation in terms of a *parametric* linear combination of predictor variables, where the parametric coefficients (intercept and parameters) are locally estimated at each location across the landscape. Formally, the parametric model at location i is in the form:

$$y(u_i, v_i) = \alpha(u_i, v_i) + \sum_{k=1}^{n} \beta_k(u_i, v_i) x_k(u_i, v_i) + \epsilon_i, \qquad (1)$$

where (u_i, v_i) represents the coordinate location of i, $\alpha(u_i, v_i)$ is the intercept at location i, $\beta_k(u_i, v_i)$ is the parameter estimate at the location i for the predictor variable x_k, $x_k(u_i, v_i)$ is the value of the k-th variable for location i, and ϵ_i is the error term. Intercept and parameter estimates are based on the assumption that observations near one another have a greater influence on each other. The weight assigned to each observation is computed on the basis of a distance decay function centered on the observation i. This decay function is modified by a bandwidth setting, that is, at which distance the weight rapidly approaches zero. The bandwidth is chosen by minimizing the Akaike Information Criteria (AIC) score [7]. The choice of the weighting scheme is a relevant step in the GWR procedure and, at this purpose, several different weighting functions are defined in the literature [2]. The more common weighting functions are Gaussian and the bi-square kernels.

Kriging [5] is a spatial statistic technique which exploits positive autocorrelation and determines a local model of the spatial phenomenon. It applies an optimal linear interpolation method to estimate unknown response values $y(u_i, v_i)$ at each location i across the landscape. $y(u_i, v_i)$ is decomposed into a structural component, which represents a mean or constant trend, a random but spatially correlated component and a random noise, which expresses measurement errors or variations inherent to the attribute of interest.

A different approach is reported in [22], where the authors present a relational regression method (Mrs-SMOTI) that builds a regression model tightly integrated with a spatial database. The method considers the geometrical representation and relative positioning of the spatial objects of different types to decide the split condition for the tree induction (e.g., towns crossed by a river and towns not crossed by any river). For the splitting decision the heuristic based on the error reduction is used. The regression problem addressed in this paper is clearly different from the task faced by Mrs-SMOTI, as we assume data which are produced at a time by a sensor network, i.e. measurements of one or more variable taken from sensors which are georeferenced through the latitude-longitude position of the measuring sensor. In any case, we have found appealing the idea of partitioning data and learning the model in a stepwise fashion at each partition to solve the problem of linear collinearity also in spatial domains. We extend this idea by proposing a segmentation of the landscape in areal units according to Boolean tests on spatial dimensions and not on predictor variables as traditional model trees do. The segmentation is tailored to identify boundaries of areal units across the landscape which group (positively) autocorrelated response values. Finally, the regression model associated to each leaf is built stepwise, but it is also synthesized to be a locally estimated regression model.

Finally, SpReCo [4] is a data mining method which addresses the spatial regression problem in a transductive learning setting and takes into account the autocorrelation of spatial data by resorting to a co-training algorithmic solution. Traditional model tree based regressors are learned from two different views of the same data. One view is defined only on original predictor variables. The other view, which accounts for the possible spatial autocorrelation, is based on aggregate variables, whose values are derived by aggregating measurements of the predictor variables in the neighborhood of each considered spatial location. According to the co-training paradigm, the model learned from a view is used to predict unlabeled data for the other during the learning process. However, only some unlabeled data are considered, namely the most reliable. The final prediction of unlabeled observations is the weighted average of the regression estimates generated by both learners. According to the transductive formulation of a regression problem, SpReCo inputs both labeled and unlabeled georeferenced data and outputs a prediction of the unlabeled ones, but no regression model is produced to predict data which are not available during the learning phase.

2.2   Transfer Learning

The major assumption in many data mining techniques is that the training and future data must be in the same feature space and have the same distribution. However, in many real world applications, this assumption may not hold. For example, in time-dependent spatial applications, such as applications of sensor network analysis, we may have to define a regression task for the domain of interest in the present (target domain), but have sufficient training data for this task available only in the past (source domain). In these cases, past data may follow a different data distribution with respect to present data and knowledge transfer would greatly improve the performance of learning, by avoiding much expensive labeling effort. In recent years, transfer learning has emerged as a new learning framework to address this kind of problem.

A survey focusing on categorizing and reviewing the current progress in transfer learning is reported in [26]. This survey revises several transfer learning techniques which are defined for different data mining tasks, including regression. Independently of the task, these techniques are classified with respect to the learning setting in which they operate. In particular, the learning setting, and consequently the transfer technique, may be inductive, transductive or semi-supervised. In the inductive setting, the target task is different from the source task, although it does not matter whether the source and target domains are the same or not. Some labeled data in the target domain are required to transfer an objective predictive model for use in the target domain. In the transductive setting, the source and target tasks are the same, while the source and target domains are different. No labeled data in the target domain are available, while labeled data in the source domain are available. Finally, in the unsupervised transfer learning setting, similar to the inductive transfer learning setting, the target task is different from but related to the source task. There are no labeled data available in either source and target domains in training. According to this categorization, the transfer learning problem we address in this paper stops halfway between the transductive transfer setting and the inductive transfer setting. As in the transductive setting, we have a unique task which admits several timestamped domains. In particular, we observe that the source timestamped domains share the same feature vector which varies across the landscape, but this spatial data distribution may drift in time [12]. On the other hand, as in the inductive transfer setting, we assume the existence of some labeled observations in the present domain (the target one), which are used to transfer the predictive models learned in the past source domains to the present ones. This transfer learning problem is related to that of transferring a knowledge from WiFi localization models across time periods and space to perform WiFi localization tasks [31,25]. Additionally, this transfer with a single task is also connected to domain adaptation which has already been investigated with similar assumptions for the knowledge transfer in text classification [9].

Although the existence of this somehow related research is documented in the literature, to the best of our knowledge our work is the first attempt to tailor a transfer technique of a purely spatially local regression model to a framework of spatio-temporal data analysis.

3   Problem Setting and Preliminary Concepts

In this Section, we formulate a definition for the regression relationship between the predictor variables and the response variable observed in a geographically distributed environment. This relationship is defined according to a field-based [29] modeling of the variables which allows us to fit ubiquity of data across the landscape. The inductive regression task is formulated to learn a definition of the regression relationship in a geographically distributed training sample. We propose to address this task by learning a piecewise definition of a space-varying (parametric) regression function which met the requirements of spatial non-stationarity posed by this task without suffering of collinearity problems. Finally, the transfer learning task is formulated to allow us to transfer regression models on a landscape across time.

3.1   Spatial Regression Definition

Formally, a *spatial regression relationship*, denoted as $\tau(U, V, Y, X_1, X_2, \ldots, X_m)$, defines the (possibly unknown) space-varying relationship between a response numeric variable Y and m predictor numeric variables X_j (with $j = 1, \ldots, m$). This relationship varies across a 2D landscape $U \times V$ (e.g. Latitude $\times$ Longitude) due to the phenomenon of spatial non-stationarity. In this formulation and according to the *field-based model*, the variation of both the response variable and the predictor variables across the landscape is mathematically defined by means of one response function $y(\cdot, \cdot)$ and m distinct predictor functions $x_j(\cdot, \cdot)$ which are respectively:

$$y \colon \mathbb{U} \times \mathbb{V} \mapsto \mathbb{Y} \qquad x_j \colon \mathbb{U} \times \mathbb{V} \mapsto \mathbb{X}_j \text{ (with } j = 1, \ldots, m), \tag{2}$$

where $\mathbb{U} \times \mathbb{V} \subseteq \mathbb{R} \times \mathbb{R}$ is the range of the Cartesian product $U \times V$; $\mathbb{Y}$ is the numeric range of response function $y(\cdot, \cdot)$ (variable Y); $\mathbb{X}_j$ is the numeric range of the predictor function $x_j(\cdot, \cdot)$ (variable X_j).

An extensional definition D of the relationship τ comprises any set of observations which are simultaneously collected across the landscape according to both the response function $(y(\cdot, \cdot))$ and the predictor functions $(x_j(\cdot, \cdot)$ with $j = 1, \ldots, m)$. The observation i of this set is the data tuple defined as follows:

$$[i, \ u_i, v_i, \ x_1(u_i, v_i), \ x_2(u_i, v_i), \ \ldots, \ x_m(u_i, v_i), \ y(u_i, v_i)], \qquad (3)$$

where i is the primary key of the data tuple one-to-one associated to the point location with coordinates (u_i, v_i). $x_j(u_i, v_i)$ is the value measured for the predictor variable X_j at the location (u_i, v_i) across the landscape, while $y(u_i, v_i)$ is the (possibly unknown) value measured for the response variable Y at (u_i, v_i). The response value $y(u_j, v_j)$ may be unknown, in this case the tuple i is unlabeled.

3.2 Spatial Regression Inductive Task

The *inductive regression task* associated to τ can be formulated as follows. Given a training data set $T \subset D$ which consists of a sample of n randomly tuples taken from D and labeled with the known values for the response variable. The goal is to learn a space-varying functional representation $f \colon \mathbb{U} \times \mathbb{V} \mapsto \mathbb{R}$ of the relationship τ such that f can be used to predict unlabeled responses at any location across the landscape. Our proposal to address this task consists of a new learner which receives training data T as input and outputs a *piecewise* definition for the space-varying function f which is defined as a geographically weighted regression tree. This tree recursively partitions the landscape surface along the spatial dimensions U and V and associates the areal unit at each leaf with a parametric (space-varying) linear combination of an opportunely chosen subset of the predictor variables. The parameters of this equation are locally estimated at each training location which falls in the leaf.

3.3 Geographically Weighted Regression Tree

Formally a *geographically weighted regression tree* f is defined as a binary tree $f = (N, E)$ where:

1. each node $n \in N$ is either an internal node or a leaf node $(N = N_I \cup N_L)$
   (a) an internal node $n \in N_I$ identifies a rectangular surface $s(n)$ over the landscape $U \times V$. The root identifies the entire landscape $U \times V$;
   (b) a leaf node $n \in N_L$ is associated with a parametric multiple linear regression function, that, for each location i falling in $s(n)$, allows the prediction of y_i according to predictor values and coefficients of the linear combination as they are locally estimated at the location i;
2. each edge $(n_i, n_j) \in E$ is a splitting edge labeled with a Boolean test over U or V which allows the identification of the rectangular surface $s(n_j) \subset s(n_i)$.

An example of a geographically weighted regression tree is reported in Figure 1.

Once the geographically weighted regression tree f is learned, it can be used to predict the response for any unlabeled observation $i' \in D$. During classification, the leaf of f which spatially contains i' is identified. The parametric function associated to this leaf is then applied to predict the unknown response value of i' by taking into account the $(u_{i'}, v_{i'})$ localization of i'.

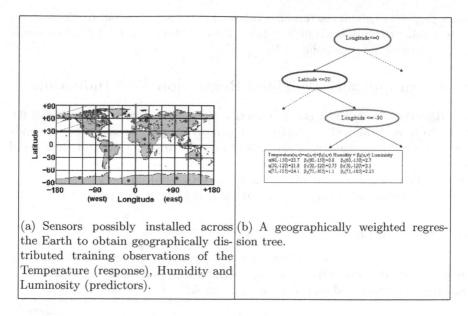

(a) Sensors possibly installed across the Earth to obtain geographically distributed training observations of the Temperature (response), Humidity and Luminosity (predictors).

(b) A geographically weighted regression tree.

Fig. 1. An example of geographically weighted regression trees with spatial splits and space-varying parametric functions at the leaves (b) learned from spatial data (a).

3.4 Transfer of Spatial Regression Models across Time

By adding the time dimension to the spatial regression task formulation, data sets are collected across the same landscape, but at distinct time points. As the manual labeling of large data sets can be very costly, it is reasonable that after an initial extensive activity of labeling, regression model(s) learned from the past would be used to predict unknown label of data at the present time point. As data distribution may drift in time, any learned model should also take this drift into account. The challenge of the transfer learning is that of allowing us to avoid that a regression model is learned again form scratch; indeed the learning operation will require to label a large amount of data to guarantee a quite accurate training. To take under control the labeling cost, only few data are labeled at the present time point and they are used to transfer regression models learned in the past across the time and adapt them to the present data at best. The hypothesis we investigate in this paper is that in presence of a set of scarcely and sparsely labeled data, we can gain more accuracy by labeling the unlabeled data with regression models transferred from the past than by using a new regression model learned from a small training set. With this aim we formulate a transfer learning task. Given, the definition of a spatial regression task $\tau(Y, X_1, X_2, \ldots, X_m, U, V)$; a series of w geographically weighted regression trees f_j, each one learned from a training source domain D_j on collected on $\mathbb{U} \times \mathbb{V}$ for the task τ at the past time t_j; and observations in the *key target set* $K \subset D$ (with responses) timestamped with the present time point. The transfer learner induces a target predictive function $f_{f_1, f_2, \ldots, f_w, K} : \mathbb{U} \times \mathbb{V} \mapsto \mathbb{Y}$ by using $f_1, f_2, \ldots f_w$ and the response values of the observations in K which allows us to predict the unlabeled observations of any *testing target set* T taken across $U \times V$ at the same time point of K.

4   Geographically Weighted Regression Tree Induction

Based on the classical Top-Down Induction of Decision Tree framework, GWRT recursively partitions the landscape in non-overlapping areal units and finds a parametric piecewise prediction model that fits training data in these areal units. Details of partitioning phase and regression model construction phase are discussed in this section. We also explain how geographically weighted regression trees can predict unknown response values across the landscape.

4.1   Splitting phase

The partitioning phase is only based on the spatial dimensions of the data. The choice of the best split is based on the well known Global Moran autocorrelation measure [18], computed on the response variable Y and defined as follows:

$$I = \frac{N}{\sum\limits_{i=1}^{N}\sum\limits_{j=1}^{N} w_{ij}} \frac{\sum\limits_{i=1}^{N}\sum\limits_{j=1}^{N} w_{ij}(y_i - \overline{y})(y_j - \overline{y})}{\sum\limits_{i=1}^{N}(y_i - \overline{y})^2}, \tag{4}$$

where y_i is the value of the response variable at the observation i, $\overline{y}$ is the mean of the response variable, w_{ij} is the spatial distance based weight between observations i and j and N is the number of observations.

The Gaussian kernel is an obvious choice to compute the weights:

$$w_{ij} = \begin{cases} e^{(-0.5d_{ij}^2/h^2)} & \text{if } d_{ij} \leq h \\ 0 & \text{otherwise} \end{cases}, \tag{5}$$

where h is the bandwidth and d_{ij} is the Euclidean spatial distance between observations i and j. The basic idea is that observations that show high positive spatial autocorrelation in the response variable should be kept in the same areal unit. Therefore, for a candidate node t, the following measure is computed:

$$I_t = \frac{(I_L N_L + I_R N_R)}{N_L + N_R} \tag{6}$$

where I_L (I_R) represents the Global Moran autocorrelation measure computed on the left (right) child of t and N_L (N_R) is the number of training observations falling in the left (right) child of t. The higher I_t, the better the split.

The candidate splits are in the form $u_i \leq \gamma_u$ or $v_i \leq \gamma_v$, where (u_i, v_i) is the spatial position of observation i. Candidate γ_u and γ_v values are determined by finding $n_{bins} - 1$ candidate equal-frequency cut points for each spatial dimension.

Our motivation of an autocorrelation measure as a splitting heuristic is that we look for a segmentation of the landscape in regions of highly correlated data which may lead to accurate GWR regression models [2].

The stopping criterion to label a node as a leaf requires the number of training observations in each node to be less than a minimum threshold. This threshold is set to the square root of the total number of training observations, which is considered a good locality threshold that does not allow too much loss in accuracy ever for rule-based classifiers [13].

4.2   Model Construction Phase

For each leaf a parametric linear regression function is associated to the areal unit associated to the leaf. Parameters of this function re estimated at each training location falling in such areal unit. After the tree is completely built, it defines a piecewice regression function f in this form $f(u, v) = \sum_{i=1}^{l} I((u, v) \in D_i) \times f_i(u, v)$, where l is the number of leaves and $D_1, D_2, \ldots, D_l$ represent the segmentation of the landscape due to the spatial partition defined by the tree; $f_i(\cdot, \cdot)$ is the parametric linear function learned for the areal unit D_i; and $I(\cdot)$ is an indicator function returning 1 if its argument is true and 0 otherwise.

Each parametric linear regression function is a parametric linear combination of a subset of the predictor variables. The variables are selected according to a forward selection strategy. Thus, the function is built with a stepwise process which starts with no variable in the function and tries out the variables one by one, including the best variable if it is "statistically significant". For each variable included in the model, parameters of the output combination are locally estimated across the landscape covered by the leaf areal unit.

To explain the stepwise construction of a parametric regression function we illustrate an example. Let us consider the case we build a function of the response variable Y with two predictor variables X_1 and X_2 and estimate the space-varying parameters of this function at the location (u_i, v_i). Our proposal is to equivalently build the parametric function:

$$\hat{y}(u_i v_i) = \alpha(u_i, v_i) + \beta(u_i, v_i)\, x_1(u_i, v_i) + \gamma(u_i, v_i)\, x_2(u_i, v_i), \qquad (7)$$

through a sequence of parametric straight-line regressions. At this aim, we start by regressing Y on X_1 and building the parametric straight line

$$\hat{y}(u_i, v_i) = \alpha_1(u_i, v_i) + \beta_1(u_i, v_i)\, x_1(u_i, v_i). \qquad (8)$$

This equation does not predict Y exactly. By adding the variable X_2, the prediction might improve. However, instead of starting from scratch and building a new function with both X_1 and X_2, we follow the stepwise procedure. First we build the parametric linear model for X_2 if X_1 is given, that is, $\hat{x}_2(u_i, v_i) = \alpha_2(u_i, v_i) + \beta_2(u_i, v_i)x_1(u_i, v_i)$. Then we compute the parametric residuals on both the predictor variable X_2 and the response variable Y, that is:

$$x_2'(u_i, v_i) = x_2(u_i, v_i) - (\alpha_2(u_i, v_i) + \beta_2(u_i, v_i)x_1(u_i, v_i)) \qquad (9)$$
$$y'(u_i, v_i) = y(u_i, v_i) - (\alpha_1(u_i, v_i) + \beta_1(u_i, v_i)x_1(u_i, v_i)). \qquad (10)$$

Finally, we determine a parametric straight-line regression between parametric residuals Y' on X_2', that is,

$$\hat{y}'(u_i, v_i) = \alpha_3(u_i, v_i) + \beta_3(u_i, v_i)x_2'(u_i, v_i). \qquad (11)$$

By substituting Equations 9-10, we reformulate Equation 11 as follows:

$$y(u_i, v_i) - (\alpha_1(u_i, v_i) + \beta_1(u_i, v_i)x_1(u_i, v_i)) = \alpha_3(u_i, v_i) + \qquad (12)$$
$$+ \beta_3(u_i, v_i)(x_2(u_i, v_i) - (\alpha_2(u_i, v_i) + \beta_2(u_i, v_i)x_1(u_i, v_i))).$$

This equation can be equivalently written as:

$$\hat{y}(u_i, v_i) = (\alpha_3(u_i, v_i) + \alpha_1(u_i, v_i) - \alpha_2(u_i, v_i)\beta_3(u_i, v_i)) + (\beta_1(u_i, v_i) - \qquad (13)$$
$$- \beta_2(u_i, v_i)\beta_3(u_i, v_i))x_1(u_i, v_i) +$$
$$+ \beta_3(u_i, v_i)x_2(u_i, v_i).$$

It can be proved that the parametric function reported in this Equation coincides with the geographically weighted model built with Y, $X1$ and $X2$ (in Equation 7) since:

$$\alpha(u_i, v_i) = \alpha_3(u_i, v_i) + \alpha_1(u_i, v_i) - \alpha_2(u_i, v_i)\beta_3(u_i, v_i), \qquad (14)$$
$$\beta(u_i, v_i) = \beta_1(u_i, v_i) - \beta_2(u_i, v_i)\beta_3(u_i, v_i) \qquad (15)$$
$$\gamma(u_i, v_i) = \beta_3(u_i, v_i). \qquad (16)$$

By considering the stepwise procedure illustrated before, two issues remain to be discussed: how parametric intercept and slope of a straight line regression (e.g. $\hat{y}(u_i, v_i) = \alpha(u_i, v_i) + \beta(u_i, v_i) \, x_j(u_i, v_i)$) are locally estimated across the landscape and how predictor variables to be added to the function are chosen.

The parametric slope and intercept are defined on the basis of the *weighted least squares regression method* [11]. This method is adapted to fit the geographically distributed arrangement of the data. In particular, for each training location which contributes to the computation of the straight-line function, the parametric slope $\beta(u_i, v_i)$ is defined as follows:

$$\beta(u_i, v_i) = (L^T \mathbf{W}_i L)^{-1} L^T \mathbf{W}_i Z, \qquad (17)$$

where L represents the vector of the values of X_j on the training observations, Z is the vector of Y values on the same observations and $\mathbf{W}_i$ is a diagonal matrix defined for the training locations (u_i, v_i) as follows: $\mathbf{W}_i = \begin{pmatrix} w_{i1} & \cdots & 0 \\ \vdots & \ddots & \vdots \\ 0 & \cdots & w_{iN} \end{pmatrix}$, where w_{ij} is computed according to Equation 5. Finally, the parametric intercept $\alpha(u_i, v_i)$ is computed according to the function:

$$\alpha(u_i, v_i) = \frac{1}{N} \sum_i z_i - \beta(u_i, v_i) \times \frac{1}{N} \sum_i l_i. \tag{18}$$

where z_i and l_i are members of Z and L, respectively.

The choice of the best predictor variable to be included in the model at each step is based on the maximization of the Akaike information criterion (AIC) measure. The AIC is a measure of the relative goodness of fitting of a statistical model. First proposed in [3], AIC is based on the concept of information entropy, and offers a relative measure of the information lost when a given model is used to describe reality. It can be said to describe the trade-off between the bias and the variance in model construction, or, loosely speaking, between the accuracy and the complexity of the model. In this work we use the corrected AIC (AIC_c) [16] that has proved to give good performance even for small datasets [7]:

$$AIC_c = 2N \, ln(\hat{\sigma}) + N \, ln(2\pi) + N \left(\frac{N+p}{N-2-p} \right), \tag{19}$$

where N is the number of training data falling in the leaf, $\hat{\sigma}$ is the standard deviation of training residuals for the response variable and p is the number of parameters (number of variables included in the model – degrees of freedom of a χ^2 test). AIC_c is used to compare regression models; however, it does not provide a test of a model in the usual sense of testing a null hypothesis; i.e. AIC can tell nothing about how well a model fits the data in an absolute sense. This means that if all the candidate models fit poorly, AIC will not give any warning. To overcome this problem, once the best variable to be added to the function is identified, the new function is evaluated according to the partial F-test. This test allows the evaluation of the statistical contribution of a new predictor variable to the model [11]. If the contribution is not statistically significant, the previous model is kept and no further variable is added to the regression function.

4.3 Prediction

Once a geographically weighted regression tree T is learned, it can be used for prediction purposes. Let o be any georeferenced observation with unknown response value, then the leaf of T spatially containing o is identified. If a training parameter estimation exists in this leaf computed for the spatial coordinates (u_o, v_o), then these estimates are used to predict $y(u_o, v_o)$ according to the local function associated to the leaf; otherwise, the k closest training parameter estimations falling in the same leaf are identified. Closeness relation is computed by the Euclidean distance. These estimated neighbor parameters are used to obtain k predictions of the response value, then a weighted combination of these responses is output. Weights are defined according to the Gaussian schema.

5   Geographically Weighted Regression Tree Transfer

Let τ be a spatial regression relationship, $f_{q-w}, f_{q-w+1}, \ldots$, and f_q be a series of w^1 functions (in our study, geographically weighted regression trees) learned for the task at the source time points $t_{q-w}, t_{q-w+1}, \ldots$, and t_q and D_{q+1} be the target domain of the task which refers to the time point t_{q+1}. The unknown response values of data collected in the time point t_{q+1} are predictable by the unknown target predictive function f_{q+1}. Then the goal is to obtain a definition of f_{q+1} by transferring the source trees towards the set of labeled key observations in the target time t_{q+1}. We denote with K_{q+1} the set of labeled target keys in t_{q+1} and T_{q+1} the set of unlabeled non-key observations.

In our proposal, a new dataset, denoted as K' is computed. It contains one tuple for each key observation in K_{q+1}. Attributes of K' represent the responses predicted by $f_{q-w}, f_{q-w+1}, \ldots$, and f_q for each key observation in K, the spatial dimension coordinates U and V and the true response value collected for the key observation. K' is now employed as training data set for a new regression task $\tau'(Y, Y_{q-w}, Y_{q-w+1}, \ldots, Y_q, U, V)$, which is formulated with the aim of learning the the target predictive function f_{q+1}.

We investigate two alternative solutions to learn f_{q+1} by a transfer learning process. The former solution employs the classical stepwise least squared regression (LSR) method [11] which uses accuracy to determine the prominent variables (responses of the past trees) for the transfer and outputs a global multivariate linear combination of these variables. The latter solution adapts the idea of a piece-wise (tree-based) form for the multivariate regression model and bases the partitioning on Boolean splits on the spatial attributes. Also in this case, the stepwise least squared regression method is used to learn the multivariate function associated to each leaf. An example of piece-wise multivariate regression model (model tree) is reported in Figure 2. Again, leaves are associated with multivariate regression models learned stepwise.

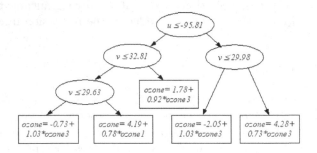

Fig. 2. An example of model tree used during transfer learning. *ozone*1 and *ozone*3 are predictor variables whose values are estimated according to f_1 and f_3, respectively.

---

[1] w is the size of a backward time window with $w \geq 1$.

6   Experiments

The inductive learner GWRT and its transfer learner, called GWRTT, are implemented in a Java system which interfaces a MySQL DBMS. In the next subsections, we illustrate results obtained with benchmark spatial data sets and a real spatio-temporal data collection.

6.1   Geographically Weighted Regression Tree Induction

GWRT is evaluated on real data collections to seek answers to the following questions. (1) How does the spatial segmentation of the landscape in rectangular areal units of positively autocorrelated responses improve both the aspatial segmentation performed by the state-of-art model tree learner M5' and the co-training solution implemented by the transductive learner SpReCo? (2) How does the stepwise construction of a piecewise space-varying parametric linear function solve the collinearity problem and improve the accuracy of traditional GWR? (3) How does the boundary bandwidth h and the neighborhood size k affect accuracy of geographically weighted regression trees induced by GWRT? In the following, we describe the data sets, the experimental setting and we illustrate results obtained with these data in order to answer questions 1-3.

Datasets. GWRT has been evaluated on six spatial regression data collections whose description is reported in the following. *Forest Fires* (FF) [8] collects 512 observations of forest fires in the period January 2000 to December 2003 in the Montesinho Natural Park, Portugal. The predictor variables are: the Fine Fuel Moisture Code, the Duff Moisture Code, the Drought Code, the Initial Spread Index, the temperature in Celsius degrees, the relative humidity, the wind speed in km/h, and the outside rain in mm/m^2. The response variable is the burned area of the forest in ha (with $1ha/100 = 100\ m^2$). The spatial coordinates (U,V) refer to the centroid of the area under investigation on a park map. *USA Geographical Analysis Spatial Data* (GASD) [24] contains 3,107 observations on USA county votes cast in the 1980 presidential election. For each county the explanatory attributes are: the population of 18 years of age or older, the population with a 12th grade or higher education, the number of owner-occupied housing units, and the aggregate income. The response attribute is the total number of votes cast. For each county, the spatial coordinates (U,V) of its centroid are available. *North-West England* (NWE)(http://www.ais.fraunhofer.de/KD/SPIN/project.html) concerns the region of North West England, which is decomposed into 1011 censual wards. Both predictor and response variables available at ward level are taken from the 1998 Census. They are the percentage of mortality (response attribute) and measures of deprivation level in the ward, according to index scores such as, Jarman Underprivileged Area Score, Townsend Score, Carstairs Score and the Department of the Environment Index. Spatial coordinates (U,V) refer to the ward centroid. By removing observations including null values, only 979 observations are used in this experiment. *Sigmea-Real* [10] collects 817 observations of the rate of herbicide resistance of two lines of plants (predictor variables), that is, the transgenic male-fertile (SMF) and the non-transgenic male-sterile (SMS) line of oilseed rape. Predictor variables are the cardinal direction and distance from the center of the donor field, the visual angle between the sampling plot and the donor field, and the shortest distance between the plot and the nearest edge of the donor field. Spatial coordinates (U,V) of the plant are available. *South California* (SC) [17] contains 8033 observations collected for the response variable, median house value, and the predictor variables, median income, housing median age, total rooms, total bedrooms, population, households in South California. Spatial coordinates represent the latitude and longitude of each observation.

Experimental Setting. The performed experiments aim at evaluating the effectiveness of the improvement of accuracy of the geographically weighted regression tree, with respect to the baseline model tree learned with the state of art model tree learner M5', the transductive learner SpReCo and the geographically weighted regression function computed by GWR. We used M5' as it is the state-of-art model tree learner which is considered as the baseline in almost all papers on model tree learning. At the best of our knowledge, no study reveals the existence of a model tree learner which definitely outperforms M5'. The implementation of M5' is publicly available in WEKA, while the implementation of GWR is publicly available in software R. M5' is run in two settings. The first setting adds the spatial dimensions to the set of predictor variables (sM5), the second setting filters out variables representing spatial dimensions (aM5). The empirical comparison between systems is based on the mean square error (MSE). To estimate the MSE, a 10-fold cross validation is performed and the average MSE (Avg.MSE) over the 10-folds is computed for each dataset. To test the significance of the difference in accuracy, we use the non-parametric Wilcoxon two-sample paired signed rank test.

Results. In Table 1, we compare the 10-fold average MSE of GWRT with the MSE of M5' (both sM5 and aM5 settings) and GWR . GWRT is run by varying h and k as we intend to draw empirical conclusions on the optimal tuning of these parameters. The results show that MSE comparison confirms that GWRT outperforms aspatial and spatial competitors, generally by a great margin. This result empirically proves the intuitions which lead us to synthesize a technique for the induction of geographically weighted regression trees. The spatial-based tree segmentation of the landscape aimed at the identification of rectangular areal units with positively autocorrelated responses improves the performance gained by the baseline M5' which partitions data (and not landscape) according to a Boolean test on the predictor variables. On the other hand, the spatial segmentation of landscape combined with linear models having a local estimate of parameters allows us to gain a more efficacious consideration (in terms of accuracy) of the spatial autocorrelation than SpReCo (which accounts the autocorrelation by a co-training procedure). The only exceptions are NWE and SMS. On the other hand, the stepwise computation of a geographically weighted regression model at each leaf is able to select the appropriate subset of

Table 1. 10-fold CV average MSE: GWRT vs M5', SpReCo and GWR. GWRT is run by varying both neighborhood size k and bandwidth h. M5' is run either by including the spatial dimensions (sM5) in the set of predictor variables or by filtering them out (aM5). GWR is run with the option for automatic bandwidth estimation. The best value of accuracy is in boldface for each dataset.

k	5	5	5	5	10	10	10	10	sM5	aM5	SpReCo	GWR
h	20%	30%	40%	50%	20%	30%	40%	50%				
FF	50.44	49.73	49.84	49.99	50.37	**49.64**	49.76	49.90	87.63	76.88	58.24	373.3
GASD	0.10	**0.09**	0.10	0.10	0.10	**0.09**	0.10	0.10	0.14	0.14	0.14	0.35
NWE	0.000	0.004	0.000	0.000	0.000	0.000	0.000	0.000	0.001	0.001	**0.0005**	0.001
SMS	11.11	5.82	4.19	4.58	18.33	5.48	4.42	4.56	3.98	4.73	**3.51**	5.22
SMF	3.66	2.24	**1.81**	1.92	3.67	2.37	1.90	1.92	2.40	1.98	1.91	1.98
SC	32.1e5	7.1e4	**5.3e4**	5.4e4	17.4e5	6.9e4	**5.3e4**	5.4e4	6.1e4	8.7e4	6.6e4	8.2e4

predictor variables at each leaf, thus solving the collinearity and definitely improving the baseline accuracy of traditional GWR. The statistical significance of the obtained differences is estimated in terms of the signed rank Wilcoxon test. The entries of Table 2 report the statistical significance of the differences between compared systems estimated with the signed rank Wilcoxon test. By insighting the statistical test results, we observe that there are three datasets, GASD, Forest Fires and South California, where GWRT statistically outperforms each competitor independently from the h and k setting (just with South California and h=20% the superiority of GWRT with respect to its competitor is not statistically obseravabe). The same primacy of GWRT is observable for the remaining three datasets, NWE, SigmeaMS and SigmeaMF, when we select higher values of h ($h >= 30\%$). In general, we observe that a choice of h between 30% and 40% leads to lower MSE in all datasets. GWRT seems to be less sensitive to the choice of k due to the weighting mechanism.

6.2   Geographically Weighted Regression Tree Transfer across Time

GWRTT is evaluated on a real spatio-temporal data collection in order to seek answers to the following questions. (1) How does the prediction function learned with the transfer technique vary in accuracy by tuning the percentage of key observations into the target domain and/or the time window size used to select source geographically weighted regression trees to be transfered across time? (2) When is the transfer learner better than the traditional inductive learner? In the following, we describe the data set, the experimental setting and we illustrate results obtained with these data in order to seek questions 1-2.

Dataset. We run experiments by considering data hourly collected by the Texas Commission On Environment Quality in the time period May 5-15, 2009. Data are obtained from 26 stations installed in Texas (http://www.tceq.state.tx.us/). Predictor variables are wind speed, temperature and solar radiation. The response variable is the ozone rate. Spatial dimensions are the latitude and longitude of the transmitting stations.

Table 2. The signed Wilcoxon test on the accuracy of systems: GWRT vs M5 (sM5 or aM5); GWRT vs SpReCo; GWRT vs GWR. The symbol "+" ("-") means that GWRT performs better (worse) than the competitor system. "*+*" ("*–*") denotes the statistically significant values ($p \leq 0.05$).

k	GWRT vs	5	5	5	5	10	10	10	10	GWRT vs	5	5	5	5	10	10	10	10
h		20	30	40	50	20	30	40	50		20	30	40	50	20	30	40	50
	sM5	+	+	+	+	+	+	+	+	aM5	+	+	+	+	+	+	+	+
FF	SpReCo	+	+	+	+	+	+	+	+	GWR	+	+	+	+	+	+	+	+
	sM5	+	+	+	+	+	+	+	+	aM5	+	+	+	+	+	+	+	+
GASD	SpReCo	+	+	+	+	+	+	+	+	GWR	+	+	+	+	+	+	+	+
NWE	sM5	-	+	+	+	-	+	+	+	aM5	-	+	+	+	-	+	+	+
	SpReCo	-	-	-	-	-	-	-	-	GWR	-	-	-	-	-	-	-	-
	sM5	-	-	+	+	-	-	+	+	aM5	-	-	+	+	-	-	+	+
SMS	SpReCo	-	-	-	-	-	-	-	-	GWR	-	+	+	+	-	+	+	+
	sM5	-	+	+	+	+	=	+	+	aM5	-	-	+	+	-	-	+	+
SMF	SpReCo	-	-	+	+	-	-	+	+	GWR	-	+	+	+	-	+	+	+
	sM5	-	+	+	+	-	+	+	+	aM5	=	+	+	+	=	+	+	+
SC	SpReCo	-	-	+	+	-	-	+	+	GWR	=	+	+	+	=	+	+	+

Experimental Settings. We define one transfer task for each day. The target domain for the transfer is the set of observations transmitted from the 26 stations at 24:00hrs of the corresponding day. The sources of the transfer are the Geographically Weighted Regression Trees learned on each source domain which is hourly collected in the corresponding day. We consider backward windows with size 1, 3, 6, 12, 18 and 24. We perform experiments by sampling 25%, 50% and 75% of the stations as key stations for the transfer.

Results. The MSE measured over the non-key observations of each target domain is averaged for the eleven days and it is illustrated in Table 3. The results of transfer are collected by varying the bandwidth h and the percentage of keys κ in the target domain. The number of neighbors k is set to 5 for all experiments. The target predictive function is induced by using either a model tree learner with Boolean test on spatial dimensions or the least square regression learner. The results of the transfer are compared with the baseline Geographically Weighted Regression Tree, induced on the key observations of the target domain only. These results suggest that, in this specific domain, LSR performs better than a model tree to approximate the target predictive function. Probably, this depends on the scarcity of key data. Additionally, we observe that the gain in accuracy due to the transfer is almost always appreciable when $\kappa = 50\%$. Concerning the bandwidth h and the window size w, we are not able to univocally identify the best setting for these parameters. This proves that further investigations are necessary in the direction of automatical parameter tuning.

Table 3. Avg MSE: eleven transfer tasks are daily defined for the Texas Ozone data collected at the 24:00 on May 5-15, 2009. For each task, sources domains are hourly collected from 0:00 to 23:00. In bold, the MSE where the transfer outperforms GWRT.

κ	w/h	Model Tree				LSR			
		20	30	40	50	20	30	40	50
25%	1	**7.58**	10.34	10.62	10.66	**4.64**	8.37	**6.17**	**6.54**
25%	3	11.78	12.31	32.60	15.84	**5.73**	**5.94**	12.87	7.91
25%	6	15.37	21.01	64.26	10.52	**6.30**	8.99	39.03	12.88
25%	12	14.55	12.38	50.30	10.79	**7.94**	**7.29**	11.55	22.46
25%	18	12.52	25.12	161.50	24.46	**7.79**	10.25	22.33	10.74
25%	24	16.33	11.29	129.89	24.37	13.77	11.34	27.30	16.73
25%	GWRT	8.03	8.08	7.85	7.59	8.03	8.08	7.85	7.59
50%	1	9.15	**7.31**	**7.31**	**6.27**	5.71	**5.93**	**5.97**	6.59
50%	3	10.99	**8.00**	10.09	**10.40**	5.49	6.51	6.00	8.12
50%	6	12.44	**10.76**	10.12	12.26	6.31	**7.03**	6.36	8.51
50%	12	11.78	**10.35**	10.60	12.73	8.76	**7.07**	**7.52**	8.29
50%	18	13.53	**11.27**	13.23	14.31	**7.22**	**7.89**	**7.94**	16.24
50%	24	17.28	**16.18**	12.96	12.94	**7.73**	**8.66**	**8.33**	**10.96**
50%	GWRT	8.33	20.24	8.67	11.10	8.33	20.24	8.67	11.10
75%	1	8.31	6.58	8.37	**6.86**	6.28	7.48	6.42	**7.28**
75%	3	10.43	7.92	11.45	15.58	**5.88**	8.31	**6.21**	**6.99**
75%	6	10.74	11.74	15.78	11.69	**5.37**	7.50	**6.09**	**7.83**
75%	12	8.69	11.67	10.57	16.29	**5.79**	7.57	**6.31**	**8.03**
75%	18	6.19	9.79	12.06	11.17	**5.19**	7.22	6.56	**7.20**
75%	24	6.67	15.90	12.15	10.92	**5.36**	8.09	6.53	**7.31**
75%	GWRT	6.08	6.08	6.42	11.45	6.08	6.08	6.42	11.45

7   Conclusions

We present a novel spatial regression technique to tackle issues posed by spatial non-stationarity and positive spatial autocorrelation in geographically distributed data environments. To deal with spatial non-stationarity we decide to learn piecewise space-varying parametric linear regression functions with coefficients which are estimated to vary across the space. We combine local model learning with a tree structured segmentation approach that recovers the functional form of a spatial model only at the level of each areal segment of the landscape. A new stepwise technique is adopted to select the most promising predictors to be included in the model, while parameters are estimated at every point across the local area. The parameter estimation solves the problem of least square weighted regression and uses a positively autocorrelated neighborhood to determine a correct estimate of the weights. Finally, a transfer learning technique is defined to transfer spatial regression models learned in the past to the present time point. Experiments with several benchmark data collections confirm that the induction of our geographically weighted regression trees generally improves both the local estimation of parameters performed by GWR and the global estimation of parameters performed by classical model tree learners like M5' as well as transductive solution for spatial regression problems as SpReCo. Furthermore, the transfer is proved to be effective in a real application. As future work, we plan to investigate techniques for automating tuning the bandwidth, the neighborhood size and the transfer window size. Additionally, we plan to use the defined transfer technique to frame this work in a streaming environment, where geographically distributed sensors continuously transmit observations across time.

Acknowledgment. This work fulfills the research objectives of both the project: "EMP3: Efficiency Monitoring of Photovoltaic Power Plants" funded by "Fondazione Cassa di Risparmio di Puglia," and the PRIN 2009 Project "Learning Techniques in Relational Domains and their Applications" funded by the Italian Ministry of University and Research (MIUR). Authors thank Lynn Rudd for her help in reading the manuscript.

References

1. Petrucci, C.S.A., Salvati, N.: The application of a spatial regression model to the analysis and mapping of poverty. Environmental and Natural Resources Series 7, 1–54 (2003)
2. Fotheringham, M.C.A.S., Brunsdon, C.: Geographically Weighted Regression: The Analysis of Spatially Varying Relationships. Wiley (2002)
3. Akaike, H.: A new look at the statistical model identification. IEEE Transactions on Automatic Control 19(6), 716–723 (1974)
4. Appice, A., Ceci, M., Malerba, D.: Transductive learning for spatial regression with co-training. In: Shin, S.Y., Ossowski, S., Schumacher, M., Palakal, M.J., Hung, C.-C. (eds.) Proceedings of the 2010 ACM Symposium on Applied Computing (SAC 2010), pp. 1065–1070. ACM (2010)
5. Bogorny, V., Valiati, J.F., da Silva Camargo, S., Engel, P.M., Kuijpers, B., Alvares, L.O.: Mining maximal generalized frequent geographic patterns with knowledge constraints. In: ICDM 2006, pp. 813–817. IEEE Computer Society (2006)
6. Brunsdon, C., McClatchey, J., Unwin, D.: Spatial variations in the average rainfall-altitude relationships in great britain: an approach using geographically weighted regression. International Journal of Climatology 21, 455–466 (2001)
7. Burnham, K., Anderson, D.: Model selection and multimodel inference: a practical information-theoretic approach. Springer (2002)
8. Cortez, P., Morais, A.: A data mining approach to predict forest fires using meteorological data. In: EPIA 2007, pp. 512–523. APPIA (2007)
9. Daumé III, H., Marcu, D.: Domain adaptation for statistical classifiers. Journal of Artificial Intelligence Research 26, 101–126 (2006)
10. Demšar, D., Debeljak, M., Lavigne, C., Džeroski, S.: Modelling pollen dispersal of genetically modified oilseed rape within the field. In: Annual Meeting of the Ecological Society of America, p. 152 (2005)
11. Draper, N.R., Smith, H.: Applied regression analysis. Wiley (1982)
12. Dries, A., Rückert, U.: Adaptive concept drift detection. Statistical Analysis and Data Mining 2(5-6), 311–327 (2009)

13. Góra, G., Wojna, A.: RIONA: A Classifier Combining Rule Induction and k-NN Method with Automated Selection of Optimal Neighbourhood. In: Elomaa, T., Mannila, H., Toivonen, H. (eds.) ECML 2002. LNCS (LNAI), vol. 2430, pp. 111–123. Springer, Heidelberg (2002)

14. Hordijk, L.: Spatial correlation in the disturbances of a linear interregional model. Regional and Urban Economics 4, 117–140 (1974)

15. Huang, Y., Leung, Y.: Analysing regional industrialisation in jiangsu province using geographically weighted regression. Journal of Geographical Systems 4, 233–249 (2002)

16. Hurvich, C.M., Tsai, C.-L.: Regression and time series model selection in small samples. Biometrika 76(2), 297–307 (1989)

17. Kelley, P., Barry, R.: Sparse spatial autoregressions. Statistics and Probability Letters 33, 291–297 (1997)

18. Legendre, P.: Spatial autocorrelation: Trouble or new paradigm? Ecology 74, 1659–1673 (1993)

19. LeSage, J., Pace, K.: Spatial dependence in data mining. In: Data Mining for Scientific and Engineering Applications, pp. 439–460. Kluwer Academic (2001)

20. Levers, C., Brückner, M., Lakes, T.: Social segregation in urban areas: an exploratory data analysis using geographically weighted regression analysis. In: 13th AGILE International Conference on Geographic Information Science 2010 (2010)

21. Longley, P., Tobon, A.: Spatial dependence and heterogeneity in patterns of hardship: an intra-urban analysis. Annals of the Association of American Geographers 94, 503–519 (2004)

22. Malerba, D., Ceci, M., Appice, A.: Mining Model Trees from Spatial Data. In: Jorge, A.M., Torgo, L., Brazdil, P.B., Camacho, R., Gama, J. (eds.) PKDD 2005. LNCS (LNAI), vol. 3721, pp. 169–180. Springer, Heidelberg (2005)

23. Mitchell, T.: Machine Learning. McGraw Hill (1997)

24. Pace, P., Barry, R.: Quick computation of regression with a spatially autoregressive dependent variable. Geographical Analysis 29(3), 232–247 (1997)

25. Pan, S.J., Shen, D., Yang, Q., Kwok, J.T.: Transferring localization models across space. In: AAAI, pp. 1383–1388. AAAI Press (2008)

26. Pan, S.J., Yang, Q.: A survey on transfer learning. IEEE Transactions on Knowledge and Data Engineering 22(10), 1345–1359 (2010)

27. Rinzivillo, S., Turini, F., Bogorny, V., Körner, C., Kuijpers, B., May, M.: Knowledge discovery from geographical data. In: Mobility, Data Mining and Privacy, pp. 243–265. Springer (2008)

28. Shariff, N., Gairola, S., Talib, A.: Modelling urban land use change using geographically weighted regression and the implications for sustainable environmental planning. In: Proceeding of the 5th International Congress on Environmental Modelling and Software Modelling for Environment's Sake, iEMSs (2010)

29. Shekhar, S., Chawla, S.: Spatial databases: A tour. Prentice Hall (2003)

30. Wang, Y., Witten, I.: Inducing Model Trees for Continuous Classes. In: van Someren, M., Widmer, G. (eds.) ECML 1997. LNCS, vol. 1224, pp. 128–137. Springer, Heidelberg (1997)

31. Zheng, V.W., Xiang, E.W., Yang, Q., Shen, D.: Transferring localization models over time. In: AAAI, pp. 1421–1426. AAAI Press (2008)

Trend Cluster Based Kriging Interpolation in Sensor Data Networks

Pietro Guccione[1], Annalisa Appice[2], Anna Ciampi[2], and Donato Malerba[2]

[1] Dipartimento di Elettrotecnica ed Elettronica, Politecnico di Bari, Bari, Italy
pietro.guccione@ieee.org
[2] Dipartimento di Informatica, Università degli Studi di Bari via Orabona,
4 - 70126 Bari, Italy
{appice,aciampi,malerba}@di.uniba.it

Abstract. Spatio-temporal data collected in sensor networks are often affected by faults due to power outage at nodes, wrong time synchronizations, interference, network transmission failures, sensor hardware issues or high energy consumption during communications. Therefore, acquisition of information by wireless sensor networks is a challenging step in monitoring physical ubiquitous phenomena (e.g. weather, pollution, traffic). This issue gives raise to a fundamental trade-off: higher density of sensors provides more data, higher resolution and better accuracy, but requires more communications and processing. A data mining approach to reduce communication and energy requirements is investigated: the number of transmitting sensors is decreased as much as possible, even keeping a reasonable degree of data accuracy. Kriging techniques and trend cluster discovery are employed to estimate unknown data in any un-sampled location of the space and at any time point of the past. Kriging is a statistical interpolation group of techniques, suited for spatial data, which estimates the unknown data in any space location by a proper weighted mean of nearby observed data. The trend clusters are stream patterns which compactly represent sensor data by means of spatial clusters having prominent data trends in time. Kriging is here applied to estimate unknown data taking into account a spatial correlation model of the sensor network. Trends are used as a guideline to transfer this model across the time horizon of the trend itself. Experiments are performed with a real sensor data network, in order to evaluate this interpolation technique and demonstrate that Kriging and trend clusters outperform, in terms of accuracy, interpolation competitors like Nearest Neighbor or Inverse Distance Weighting.

1 Introduction

Wireless sensor networks are emerging as new fundamental tools to monitor ubiquitous physical phenomena (e.g. weather, pollution, traffic). Each sensor node has a computing and storage ability and its longevity depends on the smart use of energy. The uncertainty of the application environments as well as the scarcity of communication and transmission bandwidth suggest adopting a

M. Atzmueller et al. (Eds.): MSM/MUSE 2011, LNAI 7472, pp. 118–137, 2012.
© Springer-Verlag Berlin Heidelberg 2012

meaningful subset of the whole network and estimating the un-sampled values from the available measures. High spatial densities of sensors are desirable to achieve high resolution and accurate estimates of the environmental conditions, but high densities also place heavy demands on bandwidth and energy consumption for communication [16]. The focus of this paper is on the estimation of unknown (un-sampled) data, while the problem of how to optimally select the number and the location of the sensors in a network [10,9,13] is left out. In particular, the lack of information can be due to at least three reasons:

1. A sensor has produced a large volume of data across time but, due to the memory device storage limits, this volume cannot be entirely stored for future analysis;
2. A sensor has not produced data as it is switched-off with intent (e.g. save energy) or since it has been damaged (missing data);
3. No sensor exists in a given location, but in any case we are interested in finding an estimate of the sensed quantity (new data) there.

Interpolation is a natural way to face issues caused by this lack of information: it allows the estimation of unknown data based on both known data and a model of data correlation. In such a formulation, it is evident how the choice of the data correlation model may influence the accuracy of interpolation. As it is a fact that sensor data are affected by their arrangement in space and time, it goes without saying that anyone seriously interested in determining this kind of data correlation model is required to account for the spatio-temporal arrangement of sensors in the network.

Interpolation has been seriously investigated in several research fields, ranging from artificial intelligence to data mining, statistics and signal processing. However, these interpolation techniques rarely account for spatial and temporal information simultaneously [8] and, generally, never in a streaming scenario. An interpolation technique, which used trend clusters to model the spatio-temporal dynamics of sensor data and entrusted the data estimation to the Inverse Distance Weighting (IDW) [12] technique, has been defined in previous work [6,5]. Data in an unknown location and in a given time point were estimated by computing the inverse-distance weighted sum of centroids belonging to the nearest trend cluster. A *trend cluster* [2,3] was discovered as a cluster of sensors which transmit data, whose temporal variation, called *trend polyline*, is similar along a time horizon. A trend polyline is a time series which is geo-referenced with the centroid location of the associated cluster.

In this paper, the investigation of the interpolation task proceeds by considering Kriging techniques [11] as the base for the interpolation phase. Once again these techniques are used to interpolate unknown data in any location of space by linearly combining known data but weight coefficients are defined by a second-order space varying statistics parameter. This parameter, called variogram, is an approximate measure of the spatial dissimilarity of the observed data. It combines the relative position of data with data correlation. As a matter of fact, Kriging is more complex than IDW, but it has the undeniable advantage of computing the best linear unbiased estimator of the correlation model, based on a stochastic model of the spatial dependence [7]. On the other hand, while spatial knowledge is the basis for Kriging interpolation, possible temporal correlation is neglected.

The major drawback to plug-in a Kriging technique in sensor network applications is the cost of computing a variogram, which scales as the cube of the number of observed data [11]. This cost is unacceptable especially in a sensor network where, due to concept drift, the variogram may change with time. This means that, since traditional Kriging techniques neglect data correlation in time, the computation to determine variogram has to be repeated at each new data transmission from the network. To face this complex computation issue, the trend cluster discovery is exploited to reduce the amount of data for the variogram computation and, at the same time, to avoid learning a new variogram from scratch at each new time point of the stream. The stream is segmented into user defined windows and trend clusters are discovered window-by-window. For each window, trend values at a time point of the window (e.g. at the central time point of the window) are sampled and geo-referenced in the centroids of the associated clusters and the variogram is computed from this reduced sample set. Then, contrary to the naive idea of computing the variogram from scratch at each time point of the stream, the model of temporal correlation in data which trends provide is used to transfer the variogram learned at a time point across all time points of the considered window.

The paper is organized as follows. The model employed to depict a sensor data stream, the interpolation task under consideration, the trend cluster definition and Kriging are illustrated in Section 2. The proposed Kriging interpolation framework, called TreCK (Trend Cluster based Kriging), is described in Section 3. Experiments are exposed in Section 4 in order to show how the greater complexity of Kriging gains higher accuracy in interpolation, with respect to simpler ways of interpolation, like IDW and Nearest Neighbor, by keeping computation cost under control. Conclusions are then drawn.

2   Problem Formulation and Basic Concepts

In this Section, the data model in a sensor network stream is described. Then, the interpolation task is formulated and trend clusters and Kriging are defined as basic concepts of the proposed algorithm. Finally, the ideas that mainly inspire the upgrade of the traditional Kriging techniques to the sensor network scenario are illustrated.

2.1   Sensor Data Network Model

The snapshot model [1] is adopted to represent data of a random field Z, which are continuously produced by a sensor network. Each datum measured by a sensor is geo-referenced with a 2D point location across space (e.g. latitude and longitude) and timestamped with a time point across time. The snapshot $D(t_i)$ is the set of geo-referenced data which are time-stamped with the time point t_i

across space. $D(t_i)$ can be expressed by means of the timestamped discretization of the random field function $Z(\cdot)$. This timestamped discretization of $Z(\cdot)$ maps the set of the 2D points K_i ($K_i \subseteq \mathbb{R}^2$) geo-referencing the sensors which have transmitted a measurement of Z in time t_i to the measured value.

A sensor network data stream D is the unbounded, continuous series of timestamped snapshots that is:

$$D = D(t_1), D(t_2), \ldots D(t_i), \ldots), \tag{1}$$

which are transmitted by a network at consecutive (equally-spaced) discrete time points.

In a count-based window model[4] of D, the stream is decomposed into consecutive windows of w snapshots arriving in series. A buffer continuously consumes snapshots as they arrive from the network. Once a window is completed in the buffer, it is analyzed and data are discarded; results of the data analysis are visualized and/or stored for future analysis.

2.2   Interpolation Task

Given a numeric random field, whose values are monitored through a sensor network, the goal of the interpolation task is to estimate (unknown) measurements of such a field in any location of the networked space and at any time point of the streaming period basing on the field data collected through the network.

The interpolation techniques designed for geospatial data analysis, such as the Nearest Neighbor, IDW or Kriging techniques, can be considered an initial naive approach to interpolate a random field. These techniques always obtain reliable data estimate in a space location by considering that, according to the Tobler's law of geography, a certain degree of spatial correlation is observable in measured data: nearby locations take on similar values and this similarity tends to grow if the sensors are closer to each other; or, equivalently, the power spectral density of the process implied by the measures is always band-limited [7]. However, basing the data estimation only on the spatial correlation feature has the drawback of leaving behind the analysis of the sensor data along the temporal dimension. The existence of a time dimension on sensor data measurements prompts a time interpretation for the spatial correlation: nearby locations, which *repeatedly* take on similar values at a specific time, tend to change with a similar trend in time. Based upon this consideration, it is developed the idea of interpolating unknown data of a field by using spatial clusters and data trend dynamics, which underly the sensor data of the field, rather than instantaneously interpolating the row measured data directly. Trend clusters are mined to describe both spatial clusters and data trend dynamics in sensor data.

2.3   Trend Clusters

A *trend cluster* is a *cluster* of spatially close sensors which transmit measurements of a numeric random field Z whose temporal variation, called *trend*

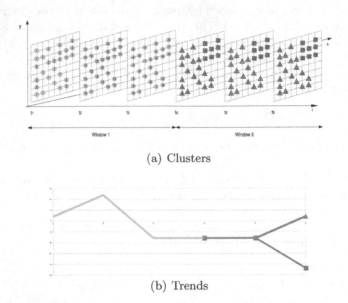

(a) Clusters

(b) Trends

Fig. 1. Trend clusters ($w = 3$). The blue cluster groups circle sensors whose measures vary as the blue polyline from t_1 to t_3. The red cluster groups square sensors whose measures vary as the red polyline from t_4 to t_6. The green cluster groups triangular sensors whose measures vary as the green polyline from t_4 to t_6.

polyline, is similar over a time horizon (see Figure 1). Formally, a trend cluster is the triple:

$$[T, c, \mathbf{z}], \tag{2}$$

where T is a *time horizon*; c is the *cluster* of spatially close sensors that measure data varying in a similar way across T; and $\mathbf{z}$ is the *trend* according to the measurements of Z streamed from c vary across T. This trend is modeled as a time series that associates each transmission time point $t \in T$ to an aggregate of the measurements of Z streamed at t by the sensors grouped in c. By performing the trend cluster discovery in a w-sized count-based model of the stream, each trend $\mathbf{z}$ is the time series of the w aggregates which are timestamped at the transmission time points of T. Formally,

$$\mathbf{z} = \langle \bar{v}_1[c], \bar{v}_2[c], \ldots, \bar{v}_{w-1}[c], \bar{v}_w[c] \rangle \tag{3}$$

where $\bar{v}_j[c]$ represents the aggregate (mean in this paper) computed on measurements clustered in c at the j-th transmission point of the window time horizon under consideration.

The spatial closeness relation, according to which the grouping of sensors in a cluster is triggered in, is a distance based nearby relation. This nearby relation is defined by the point coordinates (e.g. latitude and longitude) of sensors and depends on a distance threshold ϵ: a sensor A is ϵ-close to a sensor B (and vice-versa) in case A is far at worst ϵ from B. It is noteworthy that ϵ^1 is a parameter of the network design phase which implicitly defines virtual edges between sensors. These edges depict the spatial structure of data correlation in nearby data coherently with the Tobler's Law of Geography and contribute to define the in-network arrangement of sensors. By accounting for the network structure of sensors, a spatial cluster c is expected to group completely edged (or equivalently spatially close) sensors measuring data varying in a similar way across T. In SUMATRA, the trend cluster discovery algorithm presented in [2], the similarity in the temporal variation is evaluated with respect to a domain-driven similarity threshold denoted by δ. Spatially close sensors are grouped in the same trend cluster only iff they transmit data which persistently differ at worst δ from the associated values in the trend polyline of the cluster. The absolute difference between a sensor datum and the trended one in the cluster is evaluated at each transmission time point of the window. A detailed description of SUMATRA algorithm to discover trend clusters in the count-based model of a sensor network stream is out of the scope of this paper, but it can be found in [2].

To complete this brief description of a trend cluster pattern, the concept of a trended centroid in a sensor data stream is now introduced. By seeing a spatial cluster as a dense region around a centroid, the cluster shape can be compactly represented by its centroid. The cluster centroid is a 2D point $(\widehat{x}_c, \widehat{y}_c)$ whose coordinates are computed as follows:

$$\widehat{x}_c = \frac{1}{\sharp c} \sum_{(x,y) \in c} x \quad \text{and} \quad \widehat{y}_c = \frac{1}{\sharp c} \sum_{(x,y) \in c} y, \tag{4}$$

where $\sharp c$ is the number of sensors clustered in c. Given a trend cluster $[T, c, \mathbf{z}]$, its trended centroid is the time series $\mathbf{z}$ geo-referenced in the cluster centroid $(\widehat{x}_c, \widehat{Y}_c)$.

2.4   Kriging

Kriging is a family of techniques to interpolate the value of a random field at an unobserved location across space, starting from a known observation of its value at nearby locations and from a second order model of the field (*variogram*). Kriging provides interpolation by a linear combination of the nearby data and it is unbiased, i.e. the mean of the residual errors tends to be zero. It has also been proved that, in the family of linear interpolation, Kriging is the Best Linear Unbiased Estimator (BLUE), since it minimizes the variance of the residual errors [11].

The statistical basis of Kriging is in the consideration that a space-distributed quantity can be modeled as a random field $Z(\cdot)$. The random field is a generalization of a stochastic process, as the underlying parameter is a function of

---

[1] The choice of ϵ is than suggested by the knowledge of the domain of the measured field (e.g. the density of the sensors in a zone or the extent of the network across the space).

multidimensional vectors. In the case of a sensor network scenario, the field is a function of two (discrete) space variables and one time variable. However, in the original formulation of Kriging, time is neglected and any random field $Z(x, y)$ is a function of the space variables only. The value in any position (x^*, y^*) is estimated as follows:

$$z(x^*, y^*) = \sum_{i=1}^{N} w_i(x^*, y^*) z(x_i, y_i),$$ (5)

where N is the number of known data collected across space and each $w_i(x^*, y^*)$ is a weight to compute such a linear combination. The weights are obtained as a solution of a system of linear equations formulated by minimizing the variance of the prediction error. Rather than using weights based on an arbitrary function of distance, as for the Inverse Distance Weighted interpolation, the weights $w_i(x, y)$ are based on the computation of a *variogram* of the random field (details are in [7]).

Formally, a variogram is an approximate measure of statistical dissimilarity within the random field, taken at the given time; the higher the variogram value, the more different the values assumed by the field, on average, for that distance. Random functions, for which closely spaced values may be quite different, will have a variogram that rises quickly from the origin; random functions for which the closely spaced values are very similar will have a variogram that rises much more slowly. Given the random field, i.e. $Z(x, y)$, and assumed the enumeration of the sensed data with progressive indexes such as i, j, and so on, the sample variogram $\gamma(h)$ is defined as half the averaged square difference between the paired data values:

$$\gamma(h) = \frac{1}{2N(h)} \sum_{(i,j)|h_{i,j} \simeq h} (z_i - z_j)^2,$$ (6)

where $N(h)$ is the number of data pairs at a distance h. Equation 6 assumes the existence of an isotropic model of the field [15,7] and a proper tolerance for the distance h. The tolerance to the distance h guarantees the consideration of an acceptable number of pairs (i, j) in the empirical evaluation of the variogram.

To reduce the effect of variability, due to the unavoidable presence of noise on the measure, several research studies [7,15,11] have argued the appropriateness of fitting a theoretical model on the sample measure. The existing models are inspired by the idea that a variogram is an approximate measure of statistical dissimilarity within the random field then they have approximately the same behavior: they start with an initial low value (the *nugget*, that is the value of $\gamma(h = 0)$), then increase and, after a given distance (the *range*) they approach to an asymptotic value (the *sill*, which is $\gamma(h \to \infty)$). Moreover, it has been proved that the sill of a variogram is also the variance of the random field [7]. An example of a sample variogram and a model fitted on it using the Least Mean Square (LMS) method is illustrated in Figure 2.

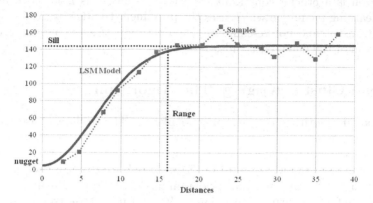

Fig. 2. An example of a sample variogram (red crosses and line) and the Gaussian model (blue line) which fits the data samples. The sill, the nugget and the range are also highlighted.

2.5 Kriging in a Sensor Network: Issues and Solutions

Adapting the traditional Kriging interpolation to a sensor network scenario is not a straightforward task. Both the issues posed by the spatial non-stationarity of a random field and the temporal non-stationarity of a variogram should be addressed and this challenge inspires the algorithm which is proposed in this paper. The analysis of both these issues is illustrated in the following.

Spatial non-stationarity of a random field. In the classical Kriging formulation, the variogram model is learned as a global function of the average squared difference within the random field, under the assumption that the function definition does not vary with space. However for random fields with large extensions the spatial invariance can no longer be supposed. More reasonably, the field is space variant and the outcomes of a sample variogram may sensibly vary on a set of data estimated in very distant locations. The idea to face this issue is to segment the surface under investigation into sub-regions, for which the invariance of the field statistics can be observed, at least up to a second order statistics like the variogram is. Based on this idea, the variogram function is conveniently computed for each one of these areas (called variogram regions).

Temporal non-stationarity of a variogram. The field measurements transmitted from each sensor are collected across time. While a variogram represents a given spatial statistics at the time of a snapshot, distinct variograms may arise in different snapshots. The temporal non-stationarity of a variogram can be naively faced by computing a new variogram at each new snapshot in the stream. By considering that the time complexity of computing a variogram is cubic in the snapshot size, the re-computation of a variogram at each new snapshot is not acceptable in a sensor network, where the assurance of a time-preserving computation is a crucial constraint. The idea to face this issue is to define a transfer learning technique which uses the prominent trends observed in data in order to transfer, and not re-compute, the definition of a variogram across the time horizon of the detected trends.

3   Trend Cluster Kriging: The Algorithm

The proposed algorithm operates in two phases. The on-line phase (see Figure 3(a)) consumes snapshots as they arrive from the sensor network, pours them, window-by-window, into TreCK framework and the data model of data is computed for the window time horizon. This model, which includes key data extracted from the window and the variogram model of data correlation, is stored in a database for any future interpolation. The off-line phase (see Figure 3(b)), which is repeatable, retrieves the model from the database and uses it for the data estimation. Details on both phases are discussed in the next subsections.

3.1   On-line Data Model Learning

The description of the interpolation model stored in database and the algorithmic details of the model computation are illustrated in the next sub-sections.

Data Model Definition. The data model $M[S,T]$ of a random field Z across the surface S and the time horizon T is the set of triples

$$M[S,T] = \{(R,D,V)\}_{R\in\mathcal{P}(S)}, \tag{7}$$

where:

1. R is a region in a partition of the surface S. This regionalization, denoted as $\mathcal{P}(S)$, segments S in non-overlapping regions, such that $\bigcup_{R\in\mathcal{P}(S)} R = S$ and $\bigcap_{R\in\mathcal{P}(S)} R = \emptyset$. Each region R is represented by means of its centroid point $(\hat{x}_R, \hat{y}_R)$ and a spatially defined membership relation $\in_{\mathcal{P}(S)}$. The centroid of a region R is computed, according to Equation 4, as the centroid of the group of sensors which fall in R. The spatial membership relation is defined as follows:

$$(x,y) \in_{\mathcal{P}(S)} R \text{ iff } R \equiv \arg\min_{R\in\mathcal{P}(S)} distance((x,y),(\hat{x}_R, \hat{y}_R)). \tag{8}$$

Given a point location (x,y), Equation 8 determines which centroid of the regionalization $\mathcal{P}(S)$ is the closest to (x,y), and establishes that (x,y) is spatially enclosed in the region of such centroid.

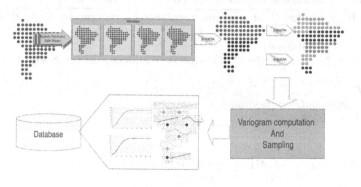

(a) Online TreCK

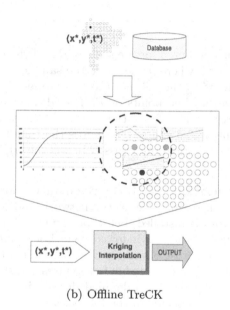

(b) Offline TreCK

Fig. 3. TreCK: on-line interpolation phase and off-line interpolation phase

2. D is a set of geo-referenced time series (polylines) with time horizon T. Formally,

$$D = \{(x, y, \mathbf{z}(x, y))\}_{(x,y)} \qquad (9)$$

where (x, y) is a point location spatially enclosed in R and $\mathbf{z}(x, y)$ is a time series geo-referenced in (x, y). For each transmission time point $t \in T$, $\mathbf{z}(x, y)$ contains the mean of the similar measurements of Z which were sensed at t by sensors located around (x, y).

Algorithm 1. function on-lineTreCK(W, DB) **return** $M[S, T]$

Require: W {a window W of w snapshots collected across surface S and time horizon T}
Require: DB {a database}
Ensure: $M[S, T]$ {data model of the window W having surface S and time horizon T}
 1: $M[S, T] \Leftarrow \oslash$
 2: $\mathcal{P}(S) \Leftarrow regionalPartitioning(W)$
 3: **for all** $R \in \mathcal{P}(S)$ **do**
 4:     $(\hat{x}_R, \hat{y}_R) \Leftarrow centroid(R)$
 5:     $\sigma^2 \Leftarrow dataVariance(R, W)$
 6:     $Sample_{(\hat{x}, \hat{y}) \in R}\{(\hat{x}, \hat{y}, \mathbf{z})\} \Leftarrow trendClusterDiscovery(R, W)$
 7:     $(\rho, \eta) \Leftarrow variogram(\{\hat{x}, \hat{y}, z_{w/2}\})$
 8:     $M[S, T] \Leftarrow add(M[S, T], ((\hat{x}_R, \hat{y}_R), Sample_{(\hat{x}, \hat{y}) \in R}\{(\hat{x}, \hat{y}, \mathbf{z})\}, (\rho, \eta, \sigma^2)))$
 9: **end for**
10: $store(M[S, T], DB)$

---

3. V is the variogram computed for the surface R and the time horizon T. It is a triple $V = (\rho, \eta, \sigma^2)$, where ρ and η are the sample range and the nugget, respectively, while σ^2 is the sill, i.e. the time series of the variance of the data field in R at the time points of T. The idea behind this formulation is that for each region $R \in \mathcal{P}(S)$ the series of timestamped variograms of R will share sample range and nugget across the time horizon T, i.e. at the time point $t \in T$, the variogram across R in t is the triple $(\rho, \eta, \sigma^2[t])$.

Data Model Computation. Let W be the data window which buffers the last w snapshots streamed from the sensor network and let S be the surface across which the network is installed. Let T be the time horizon of the window W which is discretized at the w windowed snapshot points. The data model $M[S, T]$ to estimate data across S and along T is computed from W, according to Algorithm 1. The model is computed in four phases and stored in the database for any future data interpolation. A detailed description of each phase of the interpolation algorithm is reported in the following.

Regional Segmentation Phase [lines 2-5 in Algorithm 1]. The regional segmentation $\mathcal{P}(S)$ is computed; for each region $R \in \mathcal{P}(S)$, the regional centroid $(\hat{x}_R, \hat{y}_R)$ and the time series σ^2 of the observed regional data variance are determined. The definition of $\mathcal{P}(S)$ depends on the set of trend clusters which are discovered in W with domain similarity threshold δ. For each discovered trend cluster, a region R is defined in $\mathcal{P}(S)$ and its regional centroid is computed as the spatial centroid (see Equation 4) of the associated trend cluster. In this phase, TreCK demands the trend cluster discovery to the system SUMATRA[2] and the choice of the similarity threshold δ to the user. The domain knowledge expertise should give aid to the user in the choice of the value of δ which can lead to determine the regions of the network for which a global variogram consistently fit the variability of the field. In addition, the presence of a prominent trend according to regional data drift along the window time provides the necessary knowledge to transfer the regional variogram computed at a time throughout the window time. Specifically, the time series σ^2 of the regional variance along the window time T computed in this phase will be used as means for transferring the variogram definition.

Data Reduction Phase [lines 3 and 6 in Algorithm 1]. In this phase, a sample of key data is synthesized from the windowed data and it remains available for any future interpolating manipulation. Kriging will estimate the unknown data by linearly combining these known data. In theory, TreCK should store the data window in addition to the computed variogram(s). However, in a sensor network, where the data volume can be very large, the idea of considering and storing the entire amount of sensor data is unacceptable. In order to face this issue, windowed data are *down-sampled*: the key data which will be considered for the variogram computation and will be stored in database are determined. This down-sampling phase will reduce the volume of known data, but it will preserve enough information on the spatial variability of the data sensed at each time of the window. In this way, a regional variogram will be computed from the reduced set of data for the region and Kriging will combine only the data of this set. The advantage of computing only a reduced set of the known data is twofold: speeding up the variogram determination and gaining memory space in the data model storage.

A possible down-sampling strategy could be the naive sampling technique, according to which only time series sensed by randomly selected sensors are kept. The random down-sampling is efficient, but it requires an a-priori definition of the size of the sample. On the other hand, a random choice does not preserve necessary knowledge on data variability across space. To save such information and reduce the amount of known data, the trended centroids of the trend clusters which can be discovered in each region of $\mathcal{P}(S)$ are selected. Trend clusters are discovered by using a regional domain similarity threshold which is significantly lower than the threshold of the regionalization phase. In particular, for a region $R \in \mathcal{P}(S)$, the regional domain similarity threshold of the trend cluster discovery in R is automatically decided by computing the *box plot* of field values falling in data segment of W which is spatially enclosed in R. The field values are depicted through five summaries: the smallest observation, the lower quartile (Q_1), the median (Q_2), the upper quartile (Q_3) and the largest observation. Given $\alpha = Q_1 - 1.5(Q_3 - Q_1)$ and $\beta = Q_3 + 1.5(Q_3 - Q_1)$, $\delta_R = 0.1(\beta - \alpha)$. The window is discarded and in its place the smaller key set of discovered trended centroids, denoted as $Sample_{(\hat{x},\hat{y}) \in R}\{(\hat{x}, \hat{y}, \mathbf{z})\}$, is maintained for subsequent computations.

Variogram Computation Phase [lines 3 and 7 in Algorithm 1]. The regional piecewise variogram is computed from scratch for each region of $\mathcal{P}(S)$ only at the central time point $t_{w/2}$ of T. For each region $R \in \mathcal{P}(S)$, first the Gaussian ideal model is used to to determine the range ρ and the nugget η of the variogram associated to region R at time $t_{w/2}$. Then, at each time point $t \in T$, the Gaussian ideal model (as in $t_{w/2}$), range ρ and nugget η are still assumed as they have been estimated by the fitting of the sample variogram in $t_{w/2}$. The regional variance in t, $\sigma^2(t)$, which was already computed in the regionalization phase (see line 5 of Algorithm 1), is used as the sill.

To compute the range and the nugget of each region $R \in \mathcal{P}(S)$, the sample variogram, denoted as $V_{Sample}(R, t_{w/2})$, is computed by means of Equation 6. This variogram is computed by considering the projection of $Sample_{(\hat{x}, \hat{y}) \in R}\{(\hat{x}, \hat{y}, \mathbf{z})\}$ on the time point $t_{w/2}$. The Least Mean Square (LMS) method is then used to fit the estimated sample variogram with the best Gaussian model. For the Gaussian model, $V_{Ideal}(h)$, the shape is provided in [7] with nugget zero, range one and sill one. The LMS fitted model, $V_{Model}(R, t_{w/2})$, is found by estimating the triple, $\{\eta, \rho, \sigma^2\}$ such that:

$$V_{Model}(R, t_{w/2}) = \sigma^2 \cdot V_{Ideal}(h; \rho) + \eta, \tag{10}$$

with:

$$\{\eta, \rho, \sigma^2\} = \underset{\{\eta, \rho, \sigma^2\}}{\arg\min} \left[\left(V_{Model}(R, t_{w/2}) - V_{Sample}(R, t_{w/2}) \right)^2 \right]. \tag{11}$$

After this estimation, only η, ρ are considered, since the sill, that in practice is the variance, was already computed as the time series regional variance over the window, $\sigma^2(t)$.

Storage Phase [lines 3, 8 and 10 in Algorithm 1]. For each region $R \in \mathcal{P}(S)$; the regional centroid $(\hat{x}_R, \hat{y}_R)$, the key sample $\{\hat{x}, \hat{y}, \mathbf{z}\}_{(x,y) \in R}$, the range ρ, the nugget η, and the time series of regional variances σ^2 are stored in the database as a model for the future interpolation of the Z field across the surface of R at a time point in T.

3.2 Off-line Data Interpolation

Let (x^*, y^*, t^*) the position and the time at which the estimation of the unknown value of the random field Z is required. The data model stored in database for the window hosting t^* is retrieved. Let T be the horizon of the window in the count-based model which hosts t^*. Let $M[S, T]$ be the data model (known data and piecewise regional variogram) of the surface S along time horizon T. The estimate $z(x^*, y^*, t^*)$ is determined in three phases as reported in Algorithm 2.

Data Retrieval Phase [lines 1-2 in Algorithm 2]. Kriging interpolation performs the linear combination of key data stored in $M[S, T]$ for the time t. In theory, such an estimate should combine all key data, but in practice nearest data contribute to determine the value at the unknown location more than the furthest. Based upon this consideration, only the key data falling in a neighb rhood of (x^*, y^*) is used for the estimate. As a neighborhood of a l cation (x^*, y^*), the sphere with center (x^*, y^*) and radius d is c nsidered. This radius represents the distance over which correlation is supposed to cut off, hence d is automatically determined by

---

Algorithm 2. function off-lineTreCK$(M[S,T], (x^*, y^*, t^*))$ **return** $\hat{z}(x^*, y^*, t^*)$

Require: $M[S,T]$ {the data model of the random field Z computed across surface S and past time horizon T}

Require: (x^*, y^*, t^*) {a space point (x^*, y^*) and a time point t^*}

Ensure: $\hat{z}(x^*, y^*, t^*)$ {the interpolated value for Z in (x^*, y^*, t^*)}

1: $d = 2\rho_{max}$

2: $D_d = sphericalSet\{(x, y, \mathbf{z}) \in \bigcup_{D \in M[S,T]} D \mid EuclideanDist((x, y), (x^*, y^*)) \leq d\}$

3: $R^* \Leftarrow \underset{R \in \mathcal{P}(S)}{\arg\, region}(x^*, y^*), \in_{\mathcal{P}(S)} R$

4: $(\rho, \eta, \sigma^2(t)) \Leftarrow variogram(R^*, t)$

5: $\mathbf{w}(\mathbf{x}^*, \mathbf{y}^*) \Leftarrow weight(\rho, \eta, \sigma^2(t^*), D_d)$

6: $\hat{z}(x^*, y^*, t^*) \Leftarrow \sum_{(x_i, y_i, z_i(t)) \in D_d} w_i(x^*, y^*) z_i(t^*)$

---

taking as distance a given percentage of the experimental range ρ stored in $M[S,T]$. In this paper $d = 2\rho_{max}$ with ρ_{max} the maximum experimental range determined with time horizon T.

Variogram Retrieval Phase [lines 3-5 in Algorithm 2]. The variogram stored in $M[S,T]$ is piecewise defined across the regionalization $\mathcal{P}(S)$. Therefore, first the region R^* of $\mathcal{P}(S)$ is identified such that $(x^*, y^*) \in_{\mathcal{P}(S)} R^*$. Then, the range ρ, the nugget η and the variance $\sigma^2(t^*)$ associated to this region for the time t^* are returned. This retrieved variogram is used to compute the weights w_i of linear estimation in the known data as reported in [7].

Kriging Estimate Phase [line 6 in Algorithm 2]. Finally, $z(x^*, y^*, t^*)$ is linearly estimated according to Equation 5 by combining key data with weights obtained in the previous phase.

4   Experimental Results

The TreCK framework is written in Java and interfaces a database managed by a MySQL DBMS. Experiments to evaluate the data model computation time and the interpolation accuracy are performed with the publicly available South American Air Temperature data stream [14].

Experiments are run on an Intel(R) Core(TM) 2 DUO CPU $E4500$ @2.20GHz with 2.0 GiB of RAM Memory, running Ubuntu Release 11.10 (oneiric) Kernel Linux $3.0.0 - 12 - generic$.

The goal of the experiments is twofold. First, the computation time and the interpolation accuracy of our Kriging solution is evaluated. Kriging is considered in two distinct formulations. The piecewise-regional variogram is computed from scratch only at the central time point of each window and the time series of variance is used for the transfer of the variogram across window time. The piecewise-regional variogram is computed from scratch at each snapshot of the window. The goal of considering Kriging in both these cases is to evaluate how much the use of trend clusters in transferring a variogram across the time speeds up the computation phase without significantly affecting the accuracy of interpolation. Then, the accuracy of Kriging is compared to the accuracy of competitors like IDW and 1NN. The goal of this comparative analysis is to evaluate how much Kriging outperforms IDW and 1NN in terms of accuracy of interpolation. This kind of comparison is stressed by also considering sensor networks which are not regularly populated. A sparsely populated network is obtained by switching-off a percentage of sensors in the data transmission phase. The model is learned from data sensed by switched-on sensors and it is used to estimate the field at both switched-off and switched-on sensors. It is matter of fact that Kriging works at the best when the variogram is computed from field data which are collected evenly and densely across space. Anyway, it is empirically proved that Kriging outperforms IDW and 1NN also in the case the number of transmitting sensors drops and their distribution on the original field becomes more and more skewed.

The sensor network data stream, evaluation measures and results of this empirical study are illustrated and commented in the next subsections.

4.1 Sensor Network Description

The *South American Air Climate data stream* [14] collects monthly-mean air temperature measurements (in °C) between 1960 and 1990 over a 0.5° by 0.5° of latitude/longitude grid of South America for a total of 6477 sensors. In the experiments, the network is obtained by virtually linking each sensor to the spatially close sensors which are located in the $1° \times 1°$ around cells of the grid. The recorded temperature values range between -7.6 and $32.9°C$.

4.2 Evaluation Measures and Experimental Setting

The performance of the framework are evaluated in terms of the computation time (in seconds) which is spent per window to get the elements of the data model (composed of trend clusters and variograms); the number per window of regions with a variogram associated; the number per window of centroid points where the known data are geo-referenced; the interpolation root mean squared error computed per snapshot. Kriging is compared to the version of IDW and 1NN techniques as they are properly tailored in [6,5] for the interpolation in a sensor network.

The setting-out of the parameters adopted for this experimental study is reported in Table 1. Window size w and trend similarity threshold δ are required to learn the data model (trend clusters and variograms) while no parameter setting-out is required in the estimation phase. As the accuracy of interpolation strongly depends on interpolating functions, choosing w and δ is a critical step.

In this study, the choice of the window size ($w = 12$) is motivated by the general knowledge of the specific spatial and temporal dynamics in the data distribution of a temperature quantity. In particular, the temperature is a quantity

Table 1. South American Air Climate network: setting-out of TreCK parameters

	Definition	Value
w	Window size in the data stream segmentation	12
δ	Trend similarity threshold according to spatial boundary of variogram regions (regionalization) is window-by window determined	$10°C$

with a yearlong periodic distribution. In the Northern Hemisphere the temperatures are higher in the summer than in the winter; the same occurs in the opposite seasons in the Southern Hemisphere. As snapshots are monthly timestamped in the considered stream, $w = 12$ is set in all the experiments in order to segment the stream in yearlong windows [2]. Further considerations are required to motivate our choice of δ. δ is the domain similarity threshold according to which TreCK discovers trend clusters, whose spatial boundaries define regions associated to a variogram. The choice $\delta = 10$ is suggested by the common sense according to a spatial variation of data which differs more than $10°C$ across space cannot be correctly modeled by a single variogram, provided that the dynamics of the data varies by about $40°$.

4.3   Results

In the following the results obtained for the goals reported previously are reported.

Kriging Computation Time and Accuracy. The data model is learned on consecutive windows of the entire network (all sensors are switched-on) then it is used to obtain an estimate of the streamed data. For the learning phase, the accuracy and the computation time of the variograms learned from scratch (TreCK^L) in each snapshot are compared to the accuracy and the computation time of the variograms learned in the central snapshot of each window and then transfered across window time (TreCK^T). The result of this comparison is described below.

The number of regions with a variogram associated is plotted in Figure 4(a), while the number of centroids where known data of the data model are geo-referenced is plotted in Figure 4(b). Both statistics are plotted window by window. These results show that the number of regions across which TreCK holds to determine a different correlation model is about ten per window; this number varies from 6 to 19 across the entire stream. Additionally, TreCK retains only 316 data points on average, against the 6477 data point produced each time by sensors. This down-sampling of data reduce the computation time of a variogram by taking a realistic picture of temperature dissimilarity across space.

---

[2] In this experiment, the window size can not be tailored on the basis of the moon calender, although the moon phases affect the tides. But, in this data stream temperature data are provided at the monthly aggregation level only.

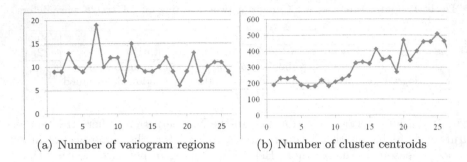

(a) Number of variogram regions   (b) Number of cluster centroids

Fig. 4. TreCK: number of regions associated to a variogram and number of centroids used as known data to learn each variogram. Both statistics are plotted per window.

Table 2. South American Air Climate Kriging (TreCK^L vs TreCK^T): average computation time (averaged per window) and average root mean squares error (averaged per snapshot)

Kriging	Average Computation Time (secs)	Average Root Mean Squares Error
TreCK^L	311.5	1.71
TreCK^T	78.5	1.94

The computation time to determine regional variograms with their trend variance and the root mean square error to estimate sensed data from the data model are plotted in Figure 5(a) and Figure 5(b). These statistics (time and error) allow the comparison between TreCK^L to TreCK^T in order to demonstrate the preserved reliability of the model computation even on this reduced set of data. Although in both solutions the computation time per window is acceptably low for the real-time data processing required by a data stream management system, it is also a fact that the TreCK^T solution gains a more remarkable reduction of time: the computation time to compute the data model, averaged per window, is cut from 311.5 (seconds) to 75.5 (seconds) as reported in Table 2.

On the other hand, as expected, the TreCK^T learning process is sped up at expense of interpolating accuracy: the analysis of the interpolation error reveals progressively worse result in each snapshot of the stream. However, this is a slight worsening since the average root mean squares error per snapshot increases slightly from 1.71 up to 1.94 (see Table 2).

Kriging vs IDw and 1NN. The accuracy of the data interpolation performed by TreCK^T is compared to the accuracy of the data interpolation performed by the IDW and 1NN data model presented in [6,5].

The capability of TreCK to accurately interpolate the field everywhere in the past is also evaluated where no field measures were collected. Hence, some experiments are performed by switching off some sensors in the network, learning the data models from switched-on sensors and using the learned model to interpo-

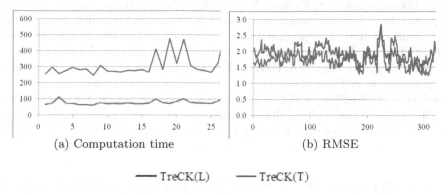

(a) Computation time   (b) RMSE

—— TreCK(L)   —— TreCK(T)

Fig. 5. TreCK: variograms learned from scratch at each snapshot (TreCK^L) vs variograms learned in the central snapshot of each window and transferred across window by using the trend variance (TreCK^T). The computation time is plotted per window, the root mean squares error is plotted per snapshot.

Table 3. South American Air Climate Interpolation (TreCK vs IDW and 1NN): average root mean squared error (averaged per snapshot)

% Switched-off Sensors	TreCK	IDW	1NN
0%	1.94	2.52	2.59
10%	2.13	2.91	2.83
20%	2.12	2.54	2.65
50%	2.08	2.67	3.01

late the entire stream. Several experimental settings are prepared by randomly switching off 0%, 10%, 20% and 50% of the sensors across the network. The root mean squared error is averaged on snapshots of the stream for TreCK, IDW and 1NN and it is reported in Table 3.

It is observable that, although Kriging is more complex than IDW and 1NN, this complexity is profitably exploited to almost halve the interpolation error when no sensor is switched-off in the network. As the primary scope of this work is to gain accuracy in data interpolation by controlling learning time for a data stream, this result is particularly valuable.

Finally, it is noteworthy that Kriging outperforms IDW and INN (see Table 3) even when a high percentage of sensors is switched-off in the network.

5   Conclusion

Trend clusters have proved to be an effective way to summarize in real-time the numeric data produced by a sensor network. Similarly, the Inverse Distance Weighting (IDW) estimation and the 1 Nearest Neighbor (1NN) estimation of such sensed data computed from the trend cluster summary stored in a database has proved to be accurate enough.

In this paper, a step forward is performed in the trend cluster definition by proposing a Kriging technique to gain accuracy in the data interpolation from trend clusters. The proposed Kriging framework, called TreCK, has been empirically evaluated on a real sensor network stream. Results show that Kriging gains accuracy with respect to IDW and 1NN and, more important, trend cluster based version of Kriging allows to control the computation time.

A future work will be addressed towards a mechanism to automatically tune the choice of the user-defined domain similarity threshold according to the networked surface is segmented into regions for the variogram computation and the size of the time horizon (window length) along which trends are discovered.

Acknowledgments. This work fulfills the research objectives of both the project: "EMP3: Efficiency Monitoring of Photovoltaic Power Plants" funded by "Fondazione Cassa di Risparmio di Puglia," and the PRIN 2009 Project "Learning Techniques in Relational Domains and their Applications" funded by the Italian Ministry of University and Research (MIUR). Authors thank Lynn Rudd for her help in reading the manuscript.

References

1. Armenakis, C.: Estimation and organization of spatio-temporal data. In: GIS (1992)
2. Ciampi, A., Appice, A., Malerba, D.: Summarization for Geographically Distributed Data Streams. In: Setchi, R., Jordanov, I., Howlett, R.J., Jain, L.C. (eds.) KES 2010. LNCS, vol. 6278, pp. 339–348. Springer, Heidelberg (2010)
3. Ciampi, A., Appice, A., Malerba, D.: Online and Offline Trend Cluster Discovery in Spatially Distributed Data Streams. In: Atzmueller, M., Hotho, A., Strohmaier, M., Chin, A. (eds.) MUSE/MSM 2010. LNCS, vol. 6904, pp. 142–161. Springer, Heidelberg (2011)
4. Gaber, M.M., Zaslavsky, A., Krishnaswamy, S.: Mining data streams: a review. ACM SIGMOD Record 34(2), 18–26 (2005)
5. Guccione, P., Ciampi, A., Appice, A., Malerba, D.: Trend cluster based interpolation everywhere in a sensor network. In: Proceedings of the 2012 ACM Symposium on Applied Computing, Data Stream, ACM SAC(DS) (2012)
6. Guccione, P., Ciampi, A., Appice, A., Malerba, D., Muolo, A.: Spatio-Temporal Reconstruction of Un-Sampled Data in a Sensor Network. In: 2nd International Workshop on Mining Ubiquitous and Social Environments (2011)
7. Isaaks, E.H., Srivastava, R.M.: An Introduction to Applied Geostatistics. Oxford University Press (1989)
8. Kerwin, W.S., Prince, J.L.: The kriging update model and recursive space-time function estimation. IEEE Transaction on Signal Processing 47(11), 2942–2952 (1999)
9. Perillo, M., Ignjatovic, Z., Heinzelman, W.: An energy conservation method for wireless sensor networks employing a blue noise spatial sampling technique. In: Information Processing in Sensor Networks, pp. 116–123 (2004)

10. Rowaihy, H., Eswaran, S., Johnson, M., Verma, D., Bar-noy, A., Brown, T.: A survey of sensor selection schemes in wireless sensor networks. In: SPIE Defense and Security Symposium Conference on Unattended Ground, Sea, and Air Sensor Technologies and Applications IX (2007)
11. Shekhar, S., Chawla, S.: The origins of kriging. Mathematical Geology 22, 239–252 (1990)
12. Shepard, D.: A two-dimensional interpolation function for irregularly-spaced data. In: Proceedings of the 23rd ACM National Conference, pp. 517–524 (1968)
13. Szczytowski, P., Khelil, A., Suri, N.: Asample: Adaptive spatial sampling in wireless sensor networks. In: SUTC/UMC, pp. 35–42 (2010)
14. S. A. A. Temperature,
    `http://climate.geog.udel.edu/c̆limate/html_pages/sa_air_clim.html`
15. Tomczak, M.: Spatial interpolation and its uncertainty using automated anisotropic inverse distance weighting (IDW) - cross-validation/jackknife approach. Journal of Geographic Information and Decision Analysis 2(2), 18–30 (1998)
16. Willett, R., Martin, A., Nowak, R.: Backcasting: A new approach to energy conservation in sensor networks. In: Information Processing in Sensor Networks, IPSN 2004 (2003)

Simulation of User Participation and Interaction in Online Discussion Groups

Else Nygren

Department of Informatics and Media,
Uppsala university, Sweden,
else.nygren@im.uu.se

Abstract. Online discussion groups (Internet forums) are difficult to analyze with normal social network analysis because there are no data that can be used to represent edges between nodes. In this study we use citations and mentions of names of other group members as a proxy for a directed social interaction between the nodes. We call these markers of social interactions *grooms*. This method: *grooming analysis* makes it possible to analyze and define a network based on the social interaction in the group. Our previous studies indicated that the tendency to make posts in the group was affected by how much grooming a group member had received from others. To test this assumption, we created various simulation models as thinking tools for understanding the mechanisms behind social structuring in discussion groups. Models were tested against observed data, with and without the concept of grooming. We found that the concept of grooming was useful to understand the mechanisms behind the activity in the group. The concept of *social grooming* - actions which invoke another participant's name, proved to be highly predictive of subsequent activity and interaction.

Keywords: Social network analysis, data mining, social interaction, online forum, Mathematica.

1 Introduction

The tendency of people to come together and form groups is inherent in the structure of society, and the ways in which such groups take shape and evolve over time is a theme that runs through large parts of social science research. [1] In an online setting, people leave digital traces that can be analyzed in a way that is generally difficult to do in the real world. One obvious type of trace is the content of the messages that people send to each other online. These can be analyzed by means of methods of content analysis. However, the content of the messages is only one aspect of the online community and in order to fully understand the activities within the group we also need to investigate the social network structure and relationships between the members of the online community. [2] Social network analysis, SNA for short, is a method that can be used to study relationships and interactions between people. SNA is different

M. Atzmueller et al. (Eds.): MSM/MUSE 2011, LNAI 7472, pp. 138–157, 2012.

from standard computer-mediated communication (CMC) methods, which often study computer-mediated relations separately from the network in which they occur. The strength of SNA lies in explaining social relations with the structure and patterns of the network in which these relations develop. [2] The use of graphical representations of the network (sociographs) helps to identify people that are central or isolated in the network, and spot asymmetries in the network structure. [3] A social network is composed of actors and ties that link the actors. In social network analysis we can describe networks of nodes connected by edges. When analyzing data from social interactions online the nodes can represent people, companies, organizations or authors. The edges in the network represent relationships between the nodes. These can be for instance friendship relations, employee relations, cooperation or co-authorship. In some social network sites we lack data to represent the edges between the nodes. This is most often the situation with online discussion groups where people post messages to the whole group but do not direct them to anyone in particular. We argue that the amount of attention given by one member of the group to another can be used as a proxy for a directed social interaction between the two, thus enabling the construction of a network representing people and the social interactions between them in an online discussion group. In this chapter we will show how we can use agent-based simulations to test various hypotheses about the factors that affect the behavior in online discussion groups. The chapter is organized as follows: first we give a background of how social networks sites currently are being studied. We then focus on how online discussion groups as a particular type of social network site can be studied and the results that have been found. Secondly we will construct simulation models of the interaction and test these against observed data from two online discussion groups. The simulation models will be described in detail. Third, we will present the simulation parameters that best mimic observed data. We will then discuss the results in the light of the findings of other researchers. Finally we will conclude and provide suggestions for further work.

2   Background

Studies of Online Sites. Studies of social online sites are interesting because they can reveal a lot about the interaction between people online and how collective behavior in groups and communities emerges. Studies have for example been made of the social network sites Flickr, Delicious, Yahoo!Answers and LinkedIn [4]; LiveJournal [5]; Wikipedia, Digg, Bugzilla and Essembly [6] as well as numerous studies related to Facebook, for instance [7]. There are also studies of the network of authors of conference publications in an online database [5], and of the networks of citations of scientific papers. [8] The most commonly used method for this kind of studies is the Social Network Analysis or SNA for short. By applying SNA to a set of gathered online data, a number of questions about social interaction online can be addressed. Some such questions center round *Membership, Growth* and *Change. Membership* questions regard what factors influence whether an individual will join or leave a group, *Growth* questions deal with what structural features influence the gain of new members in a group and *Change* questions regard how the topics of interest change over time and how that is related to the set of members in the group. At the individual level, the analysis can for example find out about the structural features characterizing the individuals position in the group, number of friends in the group and the related probability that an individual will enter or leave a group. On the group level, the analysis can find out structural features of the group like the connectedness in the form of a graph and also statistical network properties like the degree distribution, diameter and clustering coefficient. [5] However SNA only captures a snapshot in time or collapse a time interval to a static graph of the network. To study the evolution over time the methods have to be complemented by analysis of the full information about individual node and edge arrival times. Leskovec and Backstrom have shown how to study at a temporal level the micro level processes that collectively leads to the macroscopic properties of the network. In their work they have studied a wide variety of network formation strategies on the microscopic (local) scale and how these give rise to different global network structures. [4]

Studies of Online Discussion Groups. One particular type of social network site is the online discussion group. An online discussion site (also called Internet forum, Bulletin board or Message board) is a form of asynchronous conferencing where people can hold conversations in the form of posted messages. The messages are at least temporarily archived in contrast to chat rooms. Depending on the settings, users can be anonymous or have to register with a nickname to be able to post. In the later case archived posts can be tied to a particular individual by the nickname used. The structure of the online discussion site is often hierarchical or tree-like. A new discussion topic is called a thread. When a new thread is started it may either persist for some time or quickly die out. If a thread persists the people that post to that thread can be said to temporarily form a group. We call this an online discussion group. The messages in a thread are generally directed to the whole group but sometimes there are posts that specifically mention or refer to individual group members. In this way social interaction is taking place in the group. Studies of online discussion groups have been made in different fields. One such is the educational field where students participation in online discussion groups to foster learning and reflection has been extensively studied by for instance Laghos and Yang. [9] [10] In the medical field online support groups for different diagnosis and syndromes have been studied. [11] Also in the field of e-democracy and politics online discussion groups have been studied [12]. There are also a number of studies in the field of market research where the focus is on how online discussion groups can affect the attitudes of consumers towards a particular brand. [13] One of the methods used to study online discussion groups is *content analysis*. In this type of analysis the contents of the individual posts are analyzed in terms of semantic meaning and language characteristics, often according to a specific coding scheme. This method was used to analyze different types of behavior among students who participated in an educational online discussion group. [14] It has also been used to study the cross-cultural influences on communication patterns in the context of a discussion group on a popular television show. [15] Content analysis can be performed with sophisticated ways of linguistic analysis of the textual content in the messages. Dino et al used linguistic inquiry and word count to analyze the communication patterns of experienced and beginners in fan-sites. [16] When linguistic methods are used it is also possible to capture the change of topics over time in discussion groups. It is possible to see how different terms propagate among the group members via the messages they post. [17] To be able to use SNA for the study of online discussion groups it is necessary to find a proxy that can be used to represent the edges between the nodes where none are explicitly present in the data. In a study of an online discussion group for elderly people SNA was used and the mentioning of names in a message was used to indicate a directed relationship between two members. By complementing this with a content analysis it was possible to relate network structure to differences in communication patterns. [2] To capture the evolution of an online discussion in time it is necessary to analyze posts on a temporal level. A temporal analysis was used to investigate how individuals experienced an asynchronous learning environment. By microanalysis of log-files it was possible to get an in-depth understanding of the strategies they used and thus to reveal individual differences in strategies in the evolving discussion structure. [18] The different types of strategies used were found to be well described by the theoretical participation taxonomy described by Knowlton 2005. [19]

Results Found Regarding Online Discussion Groups. One result of the research on online discussion groups is that in general the level of activity (the number of posts made) varies considerably among the members in the group. When people interact with each other in online discussion groups a few people make many posts and others make only a few posts. In fact the distribution of posts among the members in the group can often be said to follow a power-law. The process by which this happens is called *preferential attachment*. This is a class of processes in which some quantity, typically some form of wealth or credit, is distributed among a number of individuals or objects according to how much they already have, so that those who are already wealthy receive more than those who are not. Under suitable circumstances, such processes generate power law distributions. [20] In fact, there are several processes that lead to this distribution and it is not clear which among them captures reality best. [4]

Disparate forms of online peer production share some common macroscopic properties that can be explained by simple dynamical mechanisms. Wilkinson have shown that user participation levels in four disparate online social sites: Wikipedia, Digg, Bugzilla and Essembly is well described by a power law, in which a few very active users account for most of the contributions [6]. According to Wilkinson the power law arise because there is a momentum associated with participation such that the probability of quitting is inversely proportional to the number of previous contributions. The power law exponent was shown to correspond clearly to the effort required to contribute, with higher exponents in systems where more effort is required. This suggests that the user participation distribution is primarily dependent on the participation momentum rule and the systems barrier to contribution. Wilkinson also showed that the distribution of contributions was lognormal because of a multiplicative reinforcement mechanism in which contributions increase popularity. This explains the propensity of a few very visible popular topics to dominate the total activity in coactive systems. Wilkinson concludes that it is rather remarkable that the many forms of variation at the individual level of these systems can be accounted for with such a simple stochastic model. [6] Wilkinson also states that, it is reasonable to assume that the outlook or philosophy of the very dedicated or prolific users will have a strong effect on the system, both by their contribution and their social interactions, which, according to Wilkinson, goes beyond any quantitative measure of prediction.

Social Grooming. Social grooming is an activity in which individuals bond and reinforce social structures. An example of social grooming is to give attention to someone. Thus to groom someone can be for instance to *mention the name of someone* in a discussion or to *cite something said by another person in the group*. By pattern string search this can be derived from records of online activity and thus analyzed numerically. This method has been described as *grooming analysis*.[21] Thus, contrary to Wilkinson, we argue that there actually is a way to quantitatively investigate such social interactions, and that is by means of using *grooming analysis* to construct the network of social interactions in the online discussion group. Our view in this chapter is that separate acts of individuals together make group properties emerge. The main acts by individuals are joining the group, making posts, interacting socially with someone and leaving the group. The collective result of these activities gives rise to network properties such as the level of activity, the growth rate of the network and the connectedness of the network. A summary will be given of the results found in a previous study including a grooming analysis of two online discussion groups. The data from that study will later be used to evaluate the simulation models. [21]

3   Summary of the Results of the Grooming Analysis in the Previous Study

The data in the study mentioned above was taken from two different international online discussion groups.

Discussion Group 1: Fashion. The first of the groups studied is the people that comment about fashion in a particular thread on the site *Fashionspot*. The website is dedicated to discussions of different aspects of design and fashion. This particular thread is about a rock band with a distinctive style in fashion. The discussions are about the style and appearance of the band members and their girlfriends but also about the music of the band. The activities that the group members engage in are viewing posts, commenting, posting pictures or links, asking questions, answering questions and generally socially chatting with each other. The data structure was as follows: Date and Time of Posting, Post Number, Member Name , Quote and a Text Field.

Discussion Group 2: Science. The second of the groups studied is the people that comment about computer science in a particular section called *The core science of simple programs* on the Internet forum *A New Kind of Science*. The discussion regards cellular automata and computational theory. The activities that the group members engage in are viewing posts, commenting, posting program files, pictures or links, asking questions, answering questions and to some extent, socially chatting with each other. The data structure was as follows: Date and Time of the Post, Member Name, and a Text Field.

In this analysis the concept of groom is defined as a social reward, in form of attention from other group members. For the Fashion group, indicator of attention was chosen to be one of two things: the first was explicit mentioning of the name of another member in a post. The second was quotation of something that another group member had written in a post. For the Science group, attention was operationalized in a similar way as one of two things: the first was explicit mentioning of the name of another member in a post. The second was commenting on something that a group member had posted. These acts were possible to identify in the data. If a group member received attention in this way, it was said that the he or she *was groomed*, or *received a groom*. For each identified act of grooming, it was registered who gave the groom and who received the groom.

A *groom balance* was constructed for each subject. This increased by one each time the subject was groomed, and was reduced by one each time the subject groomed someone else. In this way we could construct a time-dependent groom-balance for each subject. Also the accumulated number of grooms received for each subject was counted.

Data was captured by copying the content from the website pages and saving as text files. The data was imported to Mathematica and a program was made to analyse the data including picking out grooms in all of the posts.

Based on the performed analyses the behavior of the discussion groups can be described as follows. There is a constant inflow of people who check out the discussion groups. Some of these people are interested enough to make a post. If they get groomed after this first post, the probability increases that they will make a second post. If not, they will most likely leave the group. In this way a small subset of the whole group is filtered out that will keep posting. Among this active group there is an inner-circle that grooms each other. In the center of this group there is an inner core of a few people that receive an extremely large proportion of the grooming.

From the study it was found some tentative conclusions about the social dynamics in the groups: the probability of posting is higher if the person have been groomed at least once, the probability of posting is proportional to the number of grooms received, the probability of being groomed is higher if the person currently have a high positive groom balance, the probability of being groomed is proportional to the number of posts already made, the probability of leaving the group is inversely proportional to the number of grooms a person have received, the probability of leaving the group is also, to a lesser degree, proportional to the time passed since the first post.

4   Method: The Simulation Model

The Method of Simulation. For this purpose an event-based model was selected. This means that time is represented as events rather than as continuous time. An event in this context is a posting made by a member of the group. The posting may or may not include a grooming of another group member.

The social dynamics of an online discussion group will thus be modeled as a program that takes total number of posts, total number of grooms and total number of group members as input and then generate the data describing the social dynamics of the group in terms of posting activity over time, distribution of posts and grooms and the groom balance over time.

The members of the group are represented as agents although in a very simplistic way. At each point in time, the agents have access to all *current* information about posts and groom counts. Based on this information they determine whether to continue posting or not and whether to groom someone or not. The decision-making by the agents is modeled as probabilistic functions of the variables: number of posts and number of grooms. Thus the simulation model can be said to be of Monte-Carlo type.

To implement the model a program was written in Mathematica.

Basis for the Model. The nodes in our model will represent the individuals who post messages in the discussion group. We will assign each individual node a number so that node 1 represent the first individual making a post, node 2 the next individual and so on.

The edges in our model will represent a directed social interaction between two individuals. This interaction consists of someone giving a groom to another individual. Thus an edge from node i to node j represent a directed social interaction by individual i towards individual j.

For the simulation we will use an event-based model. The events are the messages that are posted in the thread in the discussion board. Therefore time proceed forward in this model, event by event. The time between two events are not simulated. The events are numbered from 1 for the first event, number 2 for the next and so on until the last event.

To reason about the simulation model we will adopt the terminology of Leskovec and Backstrom. A complete model of network evolution should incorporate the node arrival process, the edge initiation process and the edge destination process. [4]

The node arrival process governs the arrival of new nodes to the network. In the case of an online discussion group that is the rate at which new members join the discussion group by making a first post. *The edge initiation process* determines for each node when it will initiate a new edge. In our setting this represent that a member of the group will groom another member by mentioning or quoting. *The edge destination selection* process finally, determines the destination of a newly initiated edge. In our setting this corresponds to which one of the members that will receive the groom.

In addition to this we will also want to model the death of a node or *the node deletion process* that stands for a node that ceases to produce new edges. In our setting this stands for an individual that stops posting messages in the discussion group, and as a consequence also stops producing new edges because no more grooming will be done by that individual any more.

The Node Arrival Process. Backstrom et al. found a large variation in the node arrival rates, ranging from exponential to sub-linear growth. They investigated two snapshots of a network and compared how nodes arrived. They found that node arrival was related to how many friends a node already had in the network. The marginal effect of having a second friend in the network was particularly strong. However, when using the number of friends already in the network as a predictor of node arrivals they found only a slight benefit compared to random guessing. Based on this they concluded that the number of friends already in the network was not sufficient to explain the node arrival process. The observed discussion groups that we are modeling here are people that have not met before or in real life. Thus the notion of friends already in the group is not feasible to use as a modeling variable.

In our simulation model we will assume that the node arrival proceeds as follows: for each event there is a certain probability P-new-node that a new node will arrive. This probability is calculated from the input data as *Total Number of Nodes* divided by *Total Number of Events*. For each event it is determined based on this probability whether a new node will arrive or not.

The Edge Initiation Process. Leskovec and Backstrom model this so that the inter-arrival rate of the edges follows a power-law with an exponential cut-off distribution. Here we want to model this based on the viewpoint of the individual nodes. Another way of modeling this is to assume that the probability of making a second post will depend on whether the node was rewarded or not for its first post. Nodes that received no grooms are less likely to post compared to nodes that did receive grooms. In our model we have to separate the edge initiation process in two steps: the first step is posting a message and the second step is to include a groom in that message. In our earlier research we found indications that the probability of posting is higher if the person have been groomed at least once. To model this, we use a flag that is initially set to zero and is set to 1 when a node is groomed for the first time. We then assign different probabilities for a node to make a post based on this flag. These different probabilities are calculated from observed data in the following way: the probability of a groomed node to make a post is calculated as *the number of groomed nodes that makes a second post* divided by *the total number of groomed nodes*, the probability of an un-groomed node to make a post is calculated as *the number of un-groomed nodes that make a second post* divided by *the total number of un-groomed nodes*. We also found indications that the probability of posting is *proportional* to the cumulative number of grooms received. This can be modeled so that the probability of making a post is proportional to the sum of all grooms received for that particular node. We thus have the following two ways to model the first step in the edge initiation process that is to make a post: Based on *the Groom Status* and based on *the Groom Sum*. The second step in the edge initiation process is determining whether the posted message will contain a groom. There is a certain probability that a post will contain a groom. This probability is calculated as *the Total Number of Grooms* divided by *the Total Number of Posts*. For each message or event it is thus determined based on this probability whether the message will contain a groom or not.

The Edge Destination Selection Process. In our model we will try some different ways of modeling the edge destination process. One way is to assume that the probability that a certain node will be selected is the same for all nodes. This can be modeled by drawing randomly from the population of nodes in the network. This can be called the Random method. Leskovec et al. assume that nodes will close the gaps in local triangles, thus a process that is independent of the degree of the node [4]. Backstrom however assume that the edge destination is proportional to the in-degree of the destination node. In general, the in-degree is the number of in-links to a node. In online discussion groups there are no in-links so that measure cannot be used. What we can use are the number of posts and the number of grooms. In analogy with Backstroms assumptions we will model the node destination selection as proportional to the number of grooms that a certain node have received so that the probability of a node to be groomed is proportional to the current sum of grooms for that node. We also want to include a model that assumes that the node destination selection as proportional to the number of posts that a certain node have made so that the probability of a node to be groomed is proportional to the current sum of posts for that node. We thus have two methods for selecting the destination node: *Groom-Sum based* and *Post-Sum based*. Wilkinson argues that prolific users probably will have a strong effect on the system. Our interpretation of this is that he means that there are differences in the population of nodes so that some are more prolific than others. We can also say that some are *more wittier* than others and that the witty nodes probably attract more grooms than others. To model this we assume a variation in *wit-value* among the population of nodes. We also assume that this value follows a normal distribution. In the simulation model each agent is assigned a randomly drawn wit-value from this distribution and the probability to be groomed is set proportional to this value. This gives us a fourth model of the process of edge destination selection method that is called *the wit-based method*. In our former study we found that the probability of being groomed is higher if the person currently have a high positive groom balance. The groom-balance is the difference between the number of grooms a node has given away and the number of grooms that a node has received. This gives yet another method in which the probability of a node to be selected is proportional to the current groom balance for that node so that a high groom-balance increase the probability of being selected as a destination for an edge. This method is called *the groom-balance based method.*

The Node Deletion Process. For each event we will assign a certain probability that one of the nodes will leave. This probability cannot be calculated from observed data but has to be assigned as a parameter that we will call the *drop-out rate.* If a node will be deleted, it has to be determined which one. In the model of Leskovec et al. the lifetime of a node is drawn from a distribution, thus the evolution of the network cannot affect this process, it is settled from the beginning. However, in our former study we found that people who had received many grooms showed a greater tendency to stay and keep posting compared to people who had only received a few grooms. We also found some examples of that people who had been posting for a long time finally quit the group. To test this we will construct two methods of node deletion: the first method is that the probability of a node to be deleted is proportional to *the inverse of the current number of grooms* for that node (to calculate this a small number $= 0.1$ had to replace zero in the sums in order to avoid singularities). , the second method is to assume that the probability of a node to be deleted is proportional to *the age of the node.* The age of the node is calculated as the event number of the current post minus the event number when the node first joined the group. We will call these methods the *Inverse-Groom Sum* based method and the *Anti-Old* based method respectively. Wilkinson found that the probability of a node to drop out was inversely proportional to the number of messages that the node has posted. We will include this assumption in our model and thus have a method where the probability of a node to be deleted is proportional to *the inverse of the current number of posts* that the node has made. We call this method *the inverse-post-sum method.* Now we have investigated the different relations that have been proposed by various researchers and how these can be represented in a simulation model. The next step will now be to implement all these different methods in the simulation and see how they reproduce the observed behavior of the two groups studied.

Structure of the Simulation Program. The main loop is set up to create a sequence of posts until the required number of posts is created. Input parameters are the total number of nodes, the total number of posts and the total number of grooms. The parameters are used to calculate probabilities so the resulting number of posts and grooms may be different from the input parameter. The number of events however is fixed. The first step is that someone starts a discussion by making the first post. Someone has to answer that by a second post or we have no group at all. The program is thus initialized with two subjects and two posts. Next there are new posts entered consecutively until the last post. Each post in the online discussion group is represented by a posting event that is described by the following variables: the post number, the node number of the posting member, a flag to show that if grooming has taken place, and if so, the node number of the groomed subject. The model keeps track of all posts so that they are numbered from 1 and so on. Similarly all agents are numbered from 1 and on according to the order in which they appear in the group For all agents it is recorded for each event the number of posts they have made, the number of grooms they have received and the number of grooms they have given away. Certain limitations had to be set. The minimum number of agents in the group before grooming could take place was set to a parameter $= 5$. Similarly a minimum number of agents before any one would leave the group was set to a parameter $= 10$. This was to avoid the group from "dying before it started to grow". For each new post in the simulation the following things must be determined: -which agent will post? -maybe a new one? -will there be any grooms given?, if so, -which agent will be groomed?, -will an agent leave the group?, if so, -which agent will leave the group? The structure of the event-generation loop can be seen in figure 1.

Decision Points and Selection Methods. In three decision points we have to pick out agents. This can be done by different methods of selection from the set of currently available agents. If no new member enters the group, we have to select the posting agent. If grooming will take place, we have to select which agent will be groomed. If some agent will leave the group we have to select which agent. The methods to perform these selections are at the heart of the simulation. The methods corresponds to the different assumptions about relations described above.

Four different methods for selection of the posting agent are implemented. One is random, that is all current agents have the same probability of posting. This method is called *the random method* of selection. The other three methods are *the groom status based method* , *the groom sum based method*, the *postsum based method* and *the wit-based method* as described earlier.

Four different methods for selection of the agent to be groomed are implemented. One is random, that is all current agents have the same probability of being groomed. This method is called *the random method* of selection. The other three methods are *the groom balanced based method, the post sum based method* and *the wit-value based method* as described earlier.

Four different methods for selection of the agent to leave the group are implemented. One is random, that is all current agents have the same probability of being groomed. This method is called *the random method* of selection. The other three methods are *the inverse groom sum based method, the anti-old method* and *the inverse post sum based method* as described earlier.

Summary of the Different Selection Methods to Implement and Investigate. We now have a set of different methods to implement and investigate. For selection of the posting agent we have four methods: *Random, GroomStatus, GroomSum* and *PostSum*.

For selection of the groomed agent: *Random, GroomBalance, PostSum* and *Wit-value*.

For selection of the agent that will leave the group: *Random, InverseGroom-Sum, Anti-Old* and *InversePostSum*.

Method Selection Switch. The selection methods were implemented in Mathematica by using random choice with weighted lists. For each selection point a method switch was constructed so that the choice of method could be used by setting of a parameter. This switch was implemented so that multiple methods could be used. The selection could for instance be set to 80% chance of using method one and 20% chance of using method 2 etc. The input data to the model was taken from the observed discussion groups and were as follows for Fashion and Science respectively: number of agents = 187, 167; Number of posts = 1604, 971; number of Grooms = 447, 816. The different probabilities of posting for groomed and ungroomed agents respectively was taken from observed data and was for Fashion 0.72 and 0.27 and for Science 0.725 and 0.477. The parameters used in the simulation were the following: minimum number of agents before grooming = 5, minimum number of agent before leaving = 10, and the probabability that someone will leave the group in each event was set = 0.085.

4.1 Methods of Evaluation

The evaluation will be made in the following way: simulations will be run with the different methods described and with combinations of these. The simulations will be used to make graphs of the distribution of posting activity over time, the distribution of posts and the distributions of grooms. We will compare these graphs to the graphs we obtained by observing the data from the two studied discussion groups mentioned earlier. In figure 1 we can see the observed data for the two groups. Graphs for the Fashion and Science groups are shown to the left and right respectively. On top is shown the *Activity Graph* that shows the number of posts over time, (the event-number (time) is shown on the x-axis, on the y-axis is the node number). The first post can be seen in the bottom left of the plot when the first node makes the first post. Under the activity graphs we can see *the Post Distribution Graph* that show the distributions of total number of posts and *the Groom Distribution Graph* that show the total number of grooms per node. They are all similar with power law like distributions. In fact the three graphs for the two different discussion groups are remarkably similar in spite of the fact that they represent groups of very different character. However there are also differences: for the Fashion group we can see that the center of mass in the activity graph shows a slight upward shift, thus the members who are active in the beginning are successively replaced by new active members. In the Science group however, we can see that there is no such shift. The members who were active in the beginning are still active at the end. Most of the new members in this group only post once. A good simulation model should reproduce the properties of *the Activity Graph*. It should give similar growth of the group over time and also show the difference between some nodes with high regular activity and many nodes that only post once. A good model should also reproduce *the Post Distribution Graph.* It should show a power-law like structure of the distribution of posts so that a few nodes have many posts and there is a long tail of nodes that only post once. The numerical value of the maximum number of posts per node should be in the observed range that is around 160 posts for Fashion and 130 for Science. Similarly, *the Groom Distribution Graph* should show a power-law like structure of the distribution and the numerical value of the maximum number of grooms per node should be in the observed range that is around 60 grooms for Fashion and 50 for Science. Thus the evaluation will be made by *eye-balling* the graphs produced with the different methods. We will also use the Kolmogorov-Smirnov test to compare the distributions. The min- and max-values of the groom-balances for the two groups will be also be used for evaluation.

5   Analysis of the Results of the Simulations

The simulation was run with all of the selection methods described and with combinations of the methods.

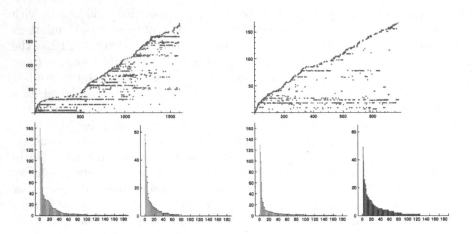

Fig. 1. OBSERVED DATA for the Fashion group left and the Science group Right. The activity graph (top) shows how the number of members grows over time and how much each member posts. Note the stratification and that many members only post once. Distribution of posts(left) and grooms(right) are shown at the bottom.

Random Methods of Selection. Fist let us ask: how well did the random based methods reproduce the observed data? The random methods correspond to mechanisms that are independent of any social network structure. People just post and groom and anyone is as likely as anyone else to post or be groomed or to drop out. Generally the random methods of selecting agents did not give graphs similar to the observed data. This was evident for two reasons: first they did not reproduce a correct distribution of posts and grooms and secondly they did not show the particular characteristics of the two groups in the activity graph. In figure 2 we can see an example of what the result of simulation looks like when all methods are set to random. The posts are evenly spread out over the time period and if we compare with the observed data in figure 1 we cannot see any traces of the typical stratification of some agents posting regularly over an extended, but limited, period of time. We can also not see that the center of mass is gradually shifted towards the right. The distributions of posts and grooms are not power law like.

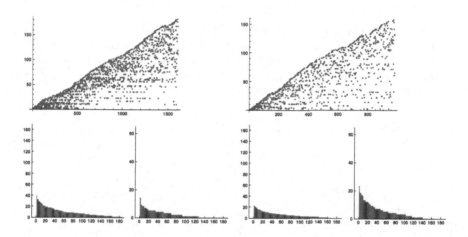

Fig. 2. RANDOM: Simulation with random methods of selection. Activity graph, distribution of posts and grooms for Fashion left and Science Right.

Table 1. Comparison of different simulation models with Kolmogorov-Smirnovs test, p-value for simulated distributions compared with observed distributions of posts and grooms for the two studied groups.

Simulation Method	Fashion Posts	Fashion Grooms	Science Posts	Science Grooms
Random	less than 0.001	less than 0.001	less than 0.001	less than 0.001
Random and Posts	less than 0.001	0.004	less than 0.001	less than 0.001
Random and AntiOld	less than 0.001	less than 0.001	less than 0.001	0.049
Random and Wit	less than 0.001	less than 0.001	less than 0.001	0.001
Posts only	0.055	0.827	less than 0.001	less than 0.001
Grooms only	less than 0.001	0.325	0.196	0.004
Posts and Grooms	0.226	0.771	0.189	0.350

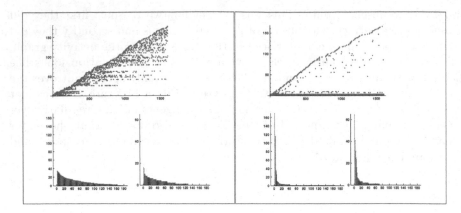

Fig. 3. RANDOM and ANTI-OLD (left), RANDOM and WIT (right. Anti-old: the probability of a node to be deleted is proportional to the age of the node, and WIT: each node is given a wit-value drawn from a normal distribution, the probability of grooming is set proportional to the wit-value. Activity graph, distribution of posts and grooms for Fashion left and Science Right.

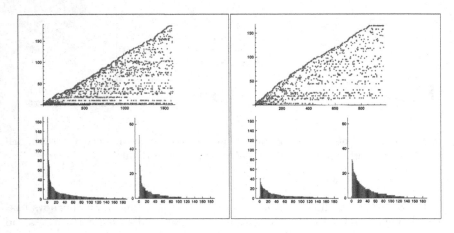

Fig. 4. POSTS ONLY. Simulation with the best methods based on random and posts. Fashion left, Science right. Activity graph, distribution of posts and grooms for Fashion left and Science Right.

Effect of the Anti-old and Wit-Value Methods. Then let us ask if the random methods could be improved, by adding anti-old method or the wit-value based method. In figure 3 we can see the results of adding either one of these two methods, the anti-old and wit-value to the left and right respectively. When the anti-old method was added we could see that the activity graph was slightly more shifted upwards. That means that it somewhat reproduced the shift that could be seen in the Fashion group. The posts and grooms however did not show

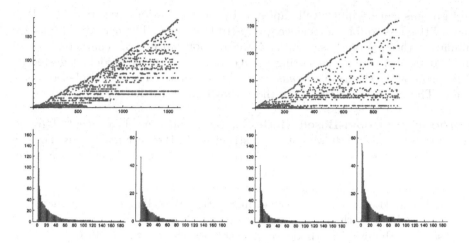

Fig. 5. POSTS AND GROOMS. Simulation with the best methods based on both number of posts and number of grooms. The activity graphs shows the charactersitic stratification and the distributions of posts and grooms are in the correct range. Fashion left, Science right. Activity graph, distribution of posts and grooms for Fashion left and Science Right.

Table 2. Comparison of different simulation models. The range of the emergent groom-balance in min and max values.

Simulation Method	Fashion Min	Fashion Max	Science Min	Science Max
Observed data	-8	15	-24	14
Random	-7	8	-8	10
Random and Posts	-10	10	-15	11
Random and AntiOld	-8	12	-10	12
Random and Wit	-9	35	-12	52
Posts only	-16	13	-8	40
Grooms only	-17	13	-24	17
Posts and Grooms	-9	17	-25	16

the observed type of distribution. By adding the wit-based method we could see a clear effect on both the graph and on the post and groom distributions. The wit-value based method made the groom distribution to be extremely skewed The effect was too strong to resemble the real data in the Science group but it is evident from the figure that this method captures some of the mechanisms in the Science group.

Groom-Free Methods of Selection. Next we will consider the random methods combined with the methods based on post counts. Let us ask how good can the models do without introducing the concept of grooms? The best combination of methods is shown in figure 4. We can see that the distribution of posts and grooms can be quite well captured by for the Fashion group (to the left) and a little less so for the Science group (to the right). The activity graphs are similar to the random based method and do not capture the characteristics of the two groups. One of the settings here corresponds to the model suggested by Wilkinson where the drop out was inversely proportional to the number of posts made. The results were similar to those shown in figure 4.

Introducing Groom-Based Methods. By introducing the concept of groom in the selection methods we got more structure in the activity graphs. For the Fashion group we got the shifting upwards over time and for the Science group we got the base of heavy posters that stay over a long time as well as many new members that posted only once as was seen in the observed data. The posts and groom distributions showed power-law like distributions but the maximum values were not as high as the observed. If the groom based methods were combined with some of the other methods we got the best fit to data as will be described below.

5.1   Best Combination of Methods for the Two Discussion Groups

No one single method reproduced the observed data for all three graphs but combinations of the methods resulted in graphs that reasonably well reproduced the observed data. In figure 5 we can see that the graphs that by the eye-balling method were deemed to be the best fit to observed data. For *the Fashion group* the observed data was best described by the simulation model when the following methods were used: -for selection of the posting agent: 1/3 chance of using the *Groom Status* method and 2/3 chance of using the *GroomSum* method. This means that the agents which have been groomed are more likely to post and that also the more you have been groomed the more likely it is that you will post. -for selection of the agent which will be groomed: 1/3 chance of using the *GroomBalance* method and 2/3 chance of using the *PostSum* method. This means that the current status of your groom balance increases your chances of getting groomed, but also that the more you have posted the more likely it is that you will be groomed. -for selection of the agent that will leave the group: 2/3 chance of using the *InverseGroomSum* method and 1/3 chance of using the *Anti Old* method. This means that the more grooms you have received the more likely it is that you will stay in the group. It also means that there is a smaller effect of length of staying in the group so that the longer you have stayed, the higher the probability that you will leave.

The *Science group* was best described by the simulation model when the following methods was used: -for selection of the posting agent: the *Groom-Sum* method. This means that the more grooms a member had received, the more likely they were to make new posts. -for selection of the agent which will be groomed: equal chances of using the *GroomBalance* method, the *PostSum* method, and the *Wit-value* method. This means that in this group both the current groom balance and the number of posts made influenced the probability of getting groomed. Also the artificial wit-value had an effect on the probability of getting groomed. -for selection of the agent that will leave the group: the *Random* method. This means that the number of grooms or the number of posts did not affect the probability of dropping out from the group.For both groups it was possible to find settings that reproduced all of the three graphs so that they showed in principle the observed behavior. In figure 6 we can also see how the groom balance over time for the Fashion group was simulated. It shows the same behavior as observed with a succession of individuals with a high groom balance. The range of the variation is also in the observed range, the same was true for the Science group.

Settings of Free Parameters. There are some free parameters in the simulation models. One of these is the *drop-out rate*, the probability for each event that a nodes will drop out, (a group member will stop posting). The sensitivity to variations in this parameter was investigated for the random methods condition. Increasing the dropout parameter successively from 1% and up showed that the number of persistent contributors successively decreased. At about 10% and over that the graphs became very unrealistic. We chose to set the dropout parameter to 8.5%. Another free parameter was the minimum number of nodes before any drop-out could take place. Setting this to one, that is no limitations, caused a quirk in the activity graph in the first events. This was because the probability of being dropped out was unreasonably high for an individual member when there were only a couple of members in the group. Increasing this value from one to about 20 removed that. A free parameter was also the minimum number of nodes before any grooming could take place. Varying this parameter from one, (that is no limitation) to 50 only reduced the total number of posts and grooms. We chose to set it to 5.

6   Conclusions

In the setting of an online discussion group, this study modeled, implemented and evaluated how nodes arrive, how edges are initiated and selected and how nodes are deleted. The limitations of this study are several. Only two relatively small groups have been used for validation.

For these two particular discussion groups we found that random methods of selecting the agents could not produce the typical power-law like distributions of posts and grooms observed.

Furthermore, addition of methods that punished old nodes, the anti-age method could reproduce certain characteristics of the activity but failed to show the correct distributions.

The methods based on counting the number of posts for each group member could display power-law-like distributions but could not account for the particular characteristics of the two groups.

The introduction of grooms in the methods, in combination with other methods could reproduce most of the particular features of the two observed groups.

The method suggested by Wilkinson, that the probability for leaving the group was inversely proportional to the number of posts made, reproduced the correct distributions of posts and grooms but failed to reproduce the particular characteristics of the activity of the two groups.

It is not just random who will post and who will leave the group. The Wilkinson model is sufficient to explain the power-law like distributions. If however one wants to investigate the particular characteristics of a certain group, then the groom-based method can provide more deep insights into the particular mechanism of that group.

A general conclusion would be that the concept of "social grooming" - actions which invoke another participant's name, are highly predictive of subsequent activity and interaction.

References

1. Coleman, J.S.: The Grid: Foundations of social theory. New ed. Belknap, Cambridge (1994)
2. Pfeil, U., Zaphiris, P.: Investigating social network patterns within an empathic online community for older people. Computers in Human Behavior 25(5), 1139–1155 (2009)
3. Scott, J.: The Grid: Social Network Analysis: A Handbook. SAGE, London (2000)
4. Leskovec, J., Backstrom, L., Kumar, R., Tomkins, A.: Microscopic Evolution of Social Networks. In: Proceedings of KDD 2008, Las Vegas, Nevada, USA, August 24-27 (2008)
5. Backstrom, L., Huttenlocher, D., Kleinberg, J., Xiangyang, L.: Group formation in large social networks: membership, growth and evolution. In: Proceedings of the 12th ACM SIGKDD International Conference on Knowledge Discovery and Data Mining. ACM, New York (2006)
6. Wilkinson, D.M.: Strong Regularities in Online Peer Production. In: Proceedings of the ACM EC 2008 Conference, Chicago, Illinois, USA, July 8-12 (2008)
7. Sheldon, K.M., Abad, N., Hinsch, C.: A Two-Process View of Facebook Use and Relatedness Need-Satisfaction: Disconnection Drives Use and Connection Rewards It. Journal of Personality and Social Psychology 100(4), 766–775 (2011)
8. Simkin, M.V., Roychowdhury, M.P.: A mathematical theory of citing. Journal of the American Society for Science and Technology 58(11), 1661–1673 (2007)
9. Laghos, A., Zaphiris, P.: Sociology of student-centred e-Learning communities: A network analysis. In: Proceedings of the IADIS International Conference, e-Society 2006, Dublin, Ireland, July 13-16 (2006)
10. Yang, X., Li, Y., Tan, C.-H., Teo, H.-H.: Students participation intention in an online discussion forum: Why is computer-mediated interaction attractive? Information and Management 44, 456–466 (2007)
11. Seale, C., Ziebland, S., Charteris-Black, J.: Gender, cancer experience and internet use: A comparative keyword analysis of interviews and online cancer support groups. Social Science and Medicine 62, 2577–2590 (2006)
12. Kim, Y.: The contribution of social network sites to exposure to political difference: The relationships among SNSs, online political messaging, and exposure to cross-cutting perspectives. Computers in Human Behavior 27, 971–977 (2011)
13. Chiou, J.-S., Lee, J.: What do they say about Friends? A cross-cultural study on Internet discussion forum. Computers in Human Behavior 24, 1179–1195 (2007)

14. Marra, R.M., Moore, J.L., Klimczak, A.K.: Content Analysis of Online Discussion Forums: A Comparative Analysis. Educational Technology Research and Development 52, 23–40 (2004)
15. Chiou, J.-S., Cheng, C.: Should a company have message boards on its web sites? Journal of Interactive Marketing 17(3), 50–61 (2003)
16. Dino, A., Reysen, S., Branscombe, N.R.: Online Interactions Between Group Members Who Differ in Status. Journal of Language and Social Psychology 28(85), 85–93 (2008)
17. Matsumura, N., Goldberg, D.E., Llora, X.: Mining Directed Social Network from Message Board. In: Proceedings of WWW 2005, Chiba, Japan, May 10-14 (2005)
18. Wise, A.F., Perera, N., Hsiao, Y.-T., Speer, J., Marbouti, F.: Microanalytic case studies of individual participation patterns in an asynchronous online discussion in an undergraduate blended course. Internet and Higher Education (2011)
19. Knowlton, D.S.: Taxonomy of learning through asynchronous discussion. Journal of Interactive Learning Research 16(2), 155–177 (2005)
20. Reka, A., Barabasi, A.-L.: Statistical mechanics of complex networks. Rev. Mod. Phys. 74, 47–97 (2002)
21. Nygren, E.: Grooming Analysis Modeling the Social Interactions of Online Discussion Groups. In: Atzmueller, M., Hotho, A., Strohmaier, M., Chin, A. (eds.) MUSE/MSM 2010. LNCS, vol. 6904, pp. 37–56. Springer, Heidelberg (2011)

Model-Driven Privacy and Security in Multi-modal Social Media UIs

Ricardo Tesoriero[1], Mohamed Bourimi[2], Fatih Karatas[2],
Thomas Barth[2], Pedro G. Villanueva[1], and Philipp Schwarte[2]

[1] Computing Systems Department,
University of Castilla-La Mancha, Albacete, Spain
{ricardo.tesoriero,pedro.gonzalez}@uclm.es
[2] IT Security Management Institute,
University of Siegen, Siegen, Germany
{bourimi,karatas,barth,schwarte}@wiwi.uni-siegen.de

Abstract. Model-driven approaches in software development are widely seen as a useful concept to a) support the formulation of non-functional requirements (NFRs) in a way domain experts are capable of, b) allow integration of multiple perspectives (from multiple domains) on the modeled system, and c) allow a stepwise refinement when actually realizing these models by a sequence of model transformations from high-, non-IT expert level down to a rather technical level. In this paper, the NFRs privacy and security are focused in the modeling of multi-modal user interfaces for social media applications. It is described how privacy and security concerns are modeled from the user interface perspective, and how this model is related to a four layer conceptual framework for developing multi-modal and multi-platform user interfaces. The approach also explains how to adapt these models to the development of social media applications. Finally, we use this proposal to model the SocialTV case study as an example of a social media application to show its feasibility.

Keywords: User interface design and modeling, non-functional requirements, usability, privacy and security, awareness and affordance, SocialTV.

1 Introduction

Due to the rapid evolution of networking technologies as well as sinking costs of communication and computing systems, the use of social networks and media is experiencing increasing usage in many important sectors of our life activities. A paradigm shift is taking place from single-user-centered usage to support multi-user needs and hence covering many collaboration forms and social aspects. Nowadays, users are interested in information exchange, collaboration and social interaction, which are supported through collaborative systems and, more recently, social software (e.g., wikis, blogs, etc.). Thereby, social media is mostly in the middle of social interaction patterns acting as collaboration catalyzer.

M. Atzmueller et al. (Eds.): MSM/MUSE 2011, LNAI 7472, pp. 158–181, 2012.

Social media is not confined to desktop computers; actually, most social applications, such as Facebook, Twitter, etc. are used on different hardware and software platforms. Due to the importance of social media in peoples life, the accessibility of social information through different modalities (vocal, gestural, tangible, etc.) is an important issue to be addressed by social media applications in order to reach disabled people. Usually, most of security and privacy concerns are boarded in terms of data access. However, from the UI perspective, there are some issues that are not related to the domain model, but they have a great impact on the UI design. The awareness and affordance are two issues that are related to the UI and are independent of the domain model of the application. For instance, resource sharing is a common situation in social media applications. Actually, some museums allow virtual and physical visitors to access multi-media resources such as, documents, music, audio, video, etc. Although from the domain model perspective, resource sharing can be reduced to a permission assignment problem; from the UI perspective, it can be solved in different ways, each of which has different consequences on awareness and affordance for users. When developing UIs for different platforms and devices, these aspects need to be taken into consideration.

To better understand the consequences of these problems for the development of UIs, let us continue with our museum example: The museum offers different kinds of resources such as pictures and texts to online as well as physical visitors equipped with a proper device. Let us now assume the museum has two different groups of resources: a) resources with which the user is authorized to interact and b) resources with which the user is not authorized to interact. From the UI perspective the central question is how to deal with these different kinds of resources.

A basic approach to solve the problem is presenting a complete list of all the resources that are available in the media library to visitors. If a visitor tries to access a resource that he/she is not allowed to access (i.e., some resources are pay per view), he/she gets an error notification. From the security perspective, this solution seems to be sufficient because the visitor is not able to access a restricted resource; from the UI perspective it is not because the visitor did not perceive any affordance or awareness about forbidden resources a priori, which yields a disappointing experience for any user. An alternative approach taking into account the affordance of the UI may only list those resources that the visitor is allowed to access, the rest of the resources remains invisible. Although this approach is the same from the security perspective; it is more attractive from the UI design perspective because visitors are not induced to make mistakes.

A more sophisticated solution may involve awareness where visitors are able to see all the resources in the media library, but there is a distinction between those that are able to be accessed from those that are not. Again from the security perspective the solution only requires a small modification that introduces the attributes that can be observed from resources that are not accessible. However form the UI perspective, this solution provides awareness and affordance. So far we have presented three different approaches to solve the same resource access problem that is inherent to the UI view of the system. As a result of this analysis we can conclude that the security and privacy issues are not only related to the domain model, they also should be addressed at the UI level.

The UsiXML [1] User Interface Description Language (UIDL) allows designers to develop multi-modal and multi-platform UIs. In this paper we propose an extension to UsiXML for addressing security and privacy issues in UI metamodeling. The approach also considers awareness and affordance of users in order to enhance user experience during interaction. First results from the application domain of SocialTV are presented.

This contribution is structured as follows: First, related work and the UsiXML UIDL are described. Afterwards, our extensions to UsiXML for modeling privacy and security issues are introduced followed by a discussion of how to model a SocialTV application utilizing these models in order to clarify the relationship between the models for the UI and the security and privacy models. A brief summary and some issues for future work conclude this contribution.

2   Related Work

Software systems and applications supporting collaboration are considered as socio-technical systems in the Computer Supported Collaborative Work (CSCW) as well as Human Computer Interaction (HCI) research fields [2,3]. The complexity of the scenarios supporting collaboration and cooperations in the respective domain is mostly reflected in the UI. Thus, HCI and CSCW also focus nowadays on human aspects (i.e., human performance and user experience) in the development of computer technology in social collaborative settings. While the goals of interaction are mostly covered by functional requirements (FRs), users' preferences (e.g., usability and awareness) and concerns (such as privacy and security) are related to NFRs. While FRs define what the system does and therefore its functionality, NFRs define how a system has to be (operational system qualities like usability, group awareness, privacy etc.) [4]. Many CSCW and HCI key literature studied therefore NFRs such as usability in socio-technical and the trade-offs, which could arise between them ,e.g., privacy and awareness trade-offs in those systems. However, various literature states that current approaches do not adequately consider generally NFRs from the beginning in the development processes such stated, e.g., for privacy and security requirements in [5, 6] and generally for all NFRs in [4].

In socio-technical systems, usability is a prerequisite for security and privacy. A major effort is therefore balancing and improving the design of security and privacy measures in the UI by considering usability aspects. However, understanding the interplay between security and usability is still one of the most disregarded and critical topics in computer security [7]. In the frame of socio-technical systems, the problems can be shown on ,e.g., UI development. From a security perspective, scattering large amounts of personal data leads to information overload, disorientation and loss of efficiency. A frequent result of this circumstance is that users do not take advantage of security options offered by applications. On the contrary, from a usability perspective the UI should address aspects such as supporting end-users in the course of balancing functionality and their individual security as well as privacy preferences with built-in mechanisms [8]. Karat et al. [7] argue, that the design of security and privacy systems yields unique aspects which differ from the many general HCI techniques. In summary, they cite that with respect to security and privacy, usability issues can have a more severe negative impact than for any other kind of systems. However, traditionally security and privacy mechanisms are implicitly designed for end-users with highly trained technical skills. The reality shows that this silent assumption does not hold and that end-users have only limited skills according to the respective scenario. Therefore end-users need facilities to easily adapt security and privacy solutions to changed requirements since they are primarily task oriented. Security and privacy are seen as complementary, but not as the main goal. Although end-users would like security and privacy mechanisms to be as transparent as possible, they also want to stay in control of the situation and understand what is going on (e.g., by means of visualizations in the UI).

While privacy is in general a desirable goal, some settings demand a certain degree of information disclosure by users. Especially in the context of social media centered settings this is necessary in order to achieve intended social collaboration goals [9]. Shneiderman et al. argue that most computer-based tasks will become collaborative because most work environments have social aspects. The authors come to this conclusion by extrapolating current trends in the field of computer-based tasks [8]. To support social interaction, different categories of collaborative and social software/hardware are required. Such solutions need to fulfill several multi-user requirements like ,e.g., social context and workspace awareness. At the same time these solutions are being characterized by the complex scenarios of the respective domain they were designed to support. Hence, the complexity of these scenarios is often reflected in the UI of these solutions [10].

Regarding privacy considering different perspectives such as CSCW, HCI, psychological and sociological aspects, a very deep analysis of the available literature can be found in [11]. There Boyle cites that users are often cautious about the way systems handle their privacy and/or security and that they are afraid that mistakes might affect their reputation. Additionally, in some circumstances users want to avoid revealing details about their current work tasks to other people. However this contradicts the concept of awareness provisioning which aims to integrate different kinds of social knowledge like the presence of people in the social media environment, their current work tasks and what he/she is doing right now [12]. This contradiction is generalized by Boyle and Greenberg in [13] and called as trade-offs between privacy and awareness. The authors describe that generally two problems are associated with providing awareness: (1) privacy violations and (2) user disruptions.

There is general consensus among researchers from the research areas cited above that security and privacy issues are caused by the way systems are designed, implemented, and deployed. But surprisingly contemporary related work does not cover modeling security and privacy as part of the early (interaction) design process for UIs. However security and usability research for developing usable (in terms of psychologically acceptable) security mechanisms is still a young research field and concrete security mechanisms are highly context-dependent [7]. Therefore we argue that related work in this field leaves room for considerable improvement on the question of how such systems can support an usable and secure user experience by means of modeling. This work contributes an answer to the following question: How can usability and security be linked together to UI elements by means of metamodels?

To answer this question conceptually adequate, linking usability, security and UI can be realized utilizing a model-driven approach [14]. Handling NFRs by model-driven approaches is already studied in several problem domains, e.g., when designing real-time systems [15] and adding NFR-specific aspects to the models or when executing scientific workflows in Grid Computing environments [16] using restructuring of a workflow to consider NFR at runtime. In contrast to these existing approaches, the central UI perspective and the complex interdependencies between various NFRs are not covered in these works. Recently many development approaches and methodologies especially in the area of sociotechnical systems follow user-centered and participatory design in combination with agile methodologies in order to efficiently react to end-users emerging needs and (change) requirements [17–20] mostly related to NFRs. However, the consideration of NFRs could be better enforced earlier by taking it as a part of the modeling step as we suggest in the following which represents the extension of our work described in [21].

3   The UsiXML UIDL Based Approach

The goal of this work is the definition of a model that copes with the introduction of security and privacy issues into the UI modeling. The best alternative for introducing these issues in a concern-independent way could be met by using the Cameleon Reference Framework (CRF) [22] to develop UIs.

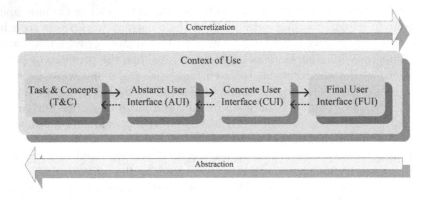

Fig. 1. Four layers of the Cameleon Reference Framework (CRF) [22]

Figure 1 presents a simplified version of the framework showing the development process divided into four steps according to the level of abstraction of the UI being addressed: **(i)** *the Tasks & Concepts (T&C)* layer describes tasks to be carried out by users, and the domain concepts required to perform these tasks, **(ii)** *the Abstract UI (AUI)* layer that defines the abstract containers and the individual components [23] that will represent the artifacts on the UI. Containers are used to group subtasks according to various criteria (e.g., task model structural patterns, cognitive load analysis, and the identification of semantic relationships). Individual component represents an artifact that describes the behavior of a UI component in a modal-independent way (navigation, task performance, etc.). Thus, the AUI model abstracts the CUI model with respect to the interaction modality, **(iii)** *the Concrete UI (CUI)* which instantiates an AUI for a given context of use. It uses Concrete Interaction Objects (CIOs) [24] to define the widget layouts and the interface navigation. It abstracts from an FUI into a UI definition that is independent of any computing platform. The CUI can also be considered as a reification of an AUI and an abstraction of the FUI with respect to the platform, and **(iv)** the Final UI (FUI) which represents the operational UI running on a particular computing platform either by interpretation (e.g., through a Web browser) or by execution (e.g., after actual code compilation).

The following set of models is defined in UsiXML supporting the conceptual modeling of UIs, and describing UIs at various abstractions levels:

- The *taskModel* that describes the task as viewed by the end user interacting with the system. It represents decomposition of task into subtasks, and temporal relationships among tasks. ConcurTaskTrees[1] (CTT) [25] can be used as a task model.

- The *domainModel* which describes the classes of objects manipulated by a user while interacting with a system. Typically, it could be a UML class diagram, or an entity-relationship-attribute model.

- The *mappingModel* that contains a series of related mappings between models, or elements of models. It gathers the inter-model relationships that are semantically related (reification, abstraction and translation).

- The *contextModel* that describes three aspects of the applications context of use where an end user is carrying out an interactive task using a specific computing platform on a surrounding environment. Thus, the context model wraps the other models to specify the deployment characteristics of the application. For instance, the user model.

- The *auiModel* which describes the UI at the abstract level as previously defined.

- The *cuiModel* that describes the UI at the concrete level as previously defined.

---

[1] http://giove.isti.cnr.it/tools/ctte/CTT_publications/publications.html

In fact, there are many UIDL candidates that follow the CRF. For instance: MariaXML[2], UIML[3], UsiXML[4], or XIML[5]. We argue that we could select any of these alternatives, but we have chosen the UsiXML [1] because it presents a set of models that can be described in terms of the MOF[6] language allowing the definition of model-to-model (M2M) transformation and model-to-text (M2T) transformations that enable the definition of a model-driven architecture (MDA) where the Security and Privacy model can be easily introduced in the architecture as new concerns. Security and privacy issues can be modeled independently from the UI and then can be related to each layer accordingly which we address in the following section.

4   Security and Privacy Models

There are various security and/or privacy models with different degrees of abstraction. However, many of these models suffer from lack of applicability in our context, which consists in supporting security and privacy requirements modeling in the UI at different abstraction levels. Most of conceptual security and privacy models are focused on technical issues/level. Because of this, we base this work on adapting the PriS approach [6] to our purposes. In the following we show how to model security and privacy concerns as well as how these models are related to the four layers of models defined by the UsiXML based approach.

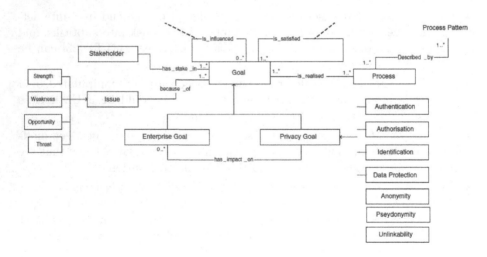

Fig. 2. PriS conceptual model (from [6])

---

[2] MariaXML: http://giove.isti.cnr.it/tools/Mariae/

[3] UIML: http://www.uiml.org

[4] UsiXML: http://www.usixml.org

[5] XIML: http://www.ximl.org

[6] The Metaobject Facility: http://www.omg.org/mof/

The PriS conceptual model (s. Fig. 2) targets earlier addressing privacy and security requirements (e.g., authentication, authorization, anonymity etc.) in the system design. This corresponds to our model-driven approach and defines security and privacy requirements for privacy enhancing technologies (PETs) and the main strength of PriS consists of addressing both, security- as well as privacy-oriented technologies at the same time. This makes it perfect as starting model for a general approach to Security and Privacy Models.

The limitation of mapping PET solutions by also considering (competing) privacy goals in the system design is addressed. Our PriS-oriented security meta-model is depicted in Fig. 3. There we took into account the most relevant security goals we need for our scenarios for now (e.g., authentication, authorization, anonymity etc.).

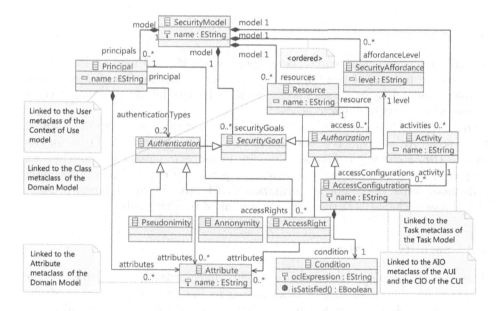

Fig. 3. PriS-oriented security metamodel

The metamodel is defined in terms of four types of entities which should represent basic entities in any security and privacy considering model in our opinion (principals, resources, security affordance and activities), and two types of relationships (access control and configuration). In the following these entities are described.

4.1 Principals

Every user in the system is assigned to a Principal that defines his/her role in the system. Therefore, a Principal defines the authentication method to be used by the users assigned to this Principal (Pseudonimity, Annonymity, Identification and so on).

Authentication methods vary according to the role played by the user; for instance, the authentication method employed by managers differs from the method employed by guests in the same system. Each Principal is linked to the user *metaclass* defined in the *context of use model* of UsiXML, which defined the user profile.

Principals define a set of *Attributes* to calculate which Resources of the system can be accessed by each of the users. An *Attribute* is characterized by a name, and it is linked to a *domain model Attribute* that provides its value. For example, in a real scenario where an Idemix server may manage the security and privacy of the system, the *Principal Attributes* define the user information that is stored into the server to grant the access to the *Resources*.

4.2 Resources

The *Resource* metaclass represents an entity of the domain model which access is restricted to a set of users defined by a *Principal*. Every *Resource* is related to one or more domain classes that define the source of information employed to define which users are able to access it.

Resources define a set of *Attributes* to constraint the accesses to the users of the system. An *Attribute* is characterized by a name, and it is linked to a *domain model Attribute* that provides its value. For instance, in a real scenario where an Idemix server may manage the security and privacy of the system, the *Resource Attributes* define the resource information that is stored into the server to grant the access to the users of the system represented by *Principals*.

4.3 Security Affordance

The *SecurityAffordance* metaclass represents different levels of security that are linked to the access of *Resources* according to the *Principal* that is accessing them. The level of affordance is represented as a sequence (or ordered collection) that defines the affordance priority; which will define how the user interface will expose this information to the user.

For instance, as stated in the Introduction section, we can expose security to the user interface in different ways: (a) we can show all resources to the user (whatever they are accessible or not), and show an error message if it is not accessible; (b) we can hide resources that are not accessible to the user; and (c) we can show accessible and not accessible resources in different ways to provide users with security awareness accordingly. Under these conditions, we define three security affordance levels: INVISIBLE, DISABLED and ENABLED. The first level has the highest priority.

Both, *AccessRights* and *AccessConfigurations*, define *securityAffordace* levels. In case these levels are in conflict when accessing a Resource, the one of the highest priority prevails.

4.4   Access Rights

The *AccessRight* metaclass represents a relationship between a *Principal* and a *Resource*, which defines the Attributes that are exposed to the security process in order to grant users (represented by a Principal) the access to *Resources*. This relationship also defines the level of security affordance for the users when accessing a *Resource*.

4.5   Activity

The *Activity* metaclass represents a set of tasks that share the same access restrictions. Each Activity is linked to a set of *tasks defined in the task model*. Thus, security issues are related to the activities performed by the system. Most of security models are focused on the resources; while they forget the context in which the resource is being accessed. This approach takes into account the task through the *Activity* concept in the security and privacy model.

4.6   Access Configuration

The *AccessConfiguration* metaclass represents different ways of accessing a *Resource* taking into account the *Activity* context. This relationship defines a condition that relates the Attributes that are exposed by the *Access Rights* to a *Security Affordance*. Thus, when a user represented by a *Principal*, is exposed to the user interface, the security and privacy model queries the *Access Configuration* relationships to calculate the *Security Affordance* of the *Activity*.

The *Access Configuration* defines the name attribute to be linked to *abstract interaction objects* (AIOs) from the *abstract user interface model* (AUI), or *concrete interaction objects* (CIOs) of the *concrete user interface model* (CUI) to describe how the user interface behavior is modified according to the security model that is being employed.

4.7   Mapping Model

The extension of the Mapping model defines the relationship among metaclasses that belong to layers at different levels of abstraction. Thus, security and privacy issues are integrated into the structure and behavior of the system. The traceability among the rest of the models is warranted by definition because of the Cameleon Conceptual Framework. The extension is depicted on Fig. 4 and exposes six intermodel relationships represented by the *InterModelRelationship* metaclass.

The relationship between the security and privacy meta-model, and the models belonging to the Task and Concepts layer of the Cameleon framework are defined by the *ActivityToTask*, *ResourceToClass* and *SecurityAttributeToDomain Attribute* intermodel relationships which represent how security issues are related to: (a) the activity being performed by the user (*ActivityToTask*), (b) the information of resources that are exposed to users, and (c) the information that is exposed to the security system.

The mapping model is also related to the Context of Use model; which defines the model of the user profile. The relationship between users and the security and privacy model is defined through the *Principal* metaclass that represents them. Thus, the *UserProfileToPrincipal* inter-model relationship defines which users are able to access the resources through the *Access Rights* relationship.

At this stage, the relationships among users, resources, security and privacy concepts have been established. However, how these relationships affect the user interface are about to be revealed. This relationship is defined at two different levels of abstraction. The key element in this relationship is the *Access Configuration* metaclass because it is related to the *abstract interaction object* (AIO) metaclass at the *abstract user interface* (AUI) layer through the *AccessConfigurationToAIO* metaclass, and to the *concrete interaction object* (CIO) metaclass at the *concrete user interface* (CUI) layer through the *AccessConfigurationToCIO* metaclass.

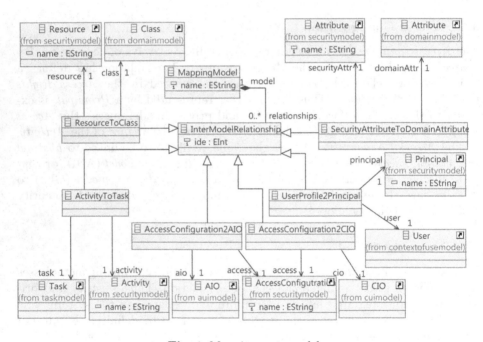

Fig. 4. Mapping metamodel

Thus, according to the *Access Configuration*, which defines the *Security Affordance* of the resources being accessed, the same *Activity* can be presented in different ways.

5   The SocialTV Case of Study

This section describes the SocialTV scenario that needs the consideration of usability as well as privacy and security issues. It also defines the domain, task, security, privacy, and abstract and concrete user interface models; and how privacy and security models are related to them.

5.1   The SocialTV Scenario

Background Information and Relevance. In an interdisciplinary SocialTV research cooperation at our University, different people from various faculties and institutes (Media Science Faculty, Marketing and IT Security and Privacy Group from the Information Systems Institute, and international HCI researchers) are working together in order to develop innovative concepts for identified requirements and needs.

Our research proceeding is mainly based on following different ethnographic approaches. For instance, by enabling people of various ages and expertise to use different SocialTV solutions in different lab and field tests. These tests are accompanied with questionnaires for different phases (e.g. two-phase evaluation with the help of one questionnaire at the beginning and another at the end). The collected data (based on usage observation and questionnaires) is later analyzed by involved researchers. In the context of this work we focus on lab and field tests carried out for selected SocialTV interaction scenarios by means of a Web based prototype to stimulate SocialTV situations. The latter primarily focused on evaluating potentials of SocialTV in the light of usability and privacy. Lab tests observations were carried out using usability testing software while field tests observations were performed in a real room allowing for a lean-back situation. In both situations, researchers are also present to observe and assist the testers. Figure 5 shows the Web based prototype used which is an online environment for supporting different collaborative SocialTV interaction scenarios. The prototype divides into four different areas (A, B, C, and D in Fig. 5). The main frame (A) shows the content on top of the chat area (B). The navigation menu (C) including the room structure was placed on left side. The general content of the prototype is arranged in rooms that can contain video material, different page types (e.g. wiki pages for text), communication channels (e.g. room-related chats and/or threaded mail) and so on. Further on this menu bar (D) contains synchronous presence awareness indicators (online users in the SocialTV platform). Awareness information is also shown in the navigation tree (C). So, users can see who else is currently in the same room watching the same video or TV content (by showing the picture of the user next to the room node in the navigation tree).

For accuracy, the used Silverlights Gorilla usability software (s. Fig. 5-E) allowed to capture the actions carried out in the SocialTV environment as well as the testers interaction in audio and video (e.g. Fig 5-F). The results of lab and field tests involving more than 32 people are presented in Figure 6. While the different colored problem areas in Figure 6) are mainly related to usability and

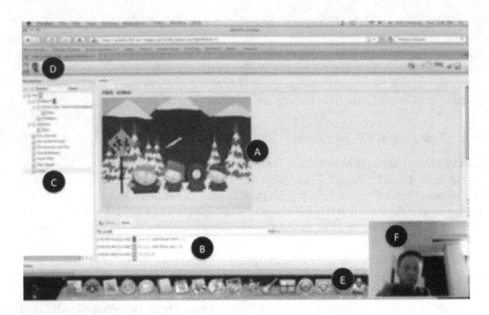

Fig. 5. Screenshot from a lab test using a SocialTV Web based prototype for synchronous and asynchronous collaboration

interaction design problems, security and privacy concerns analysis were identified with the help of the questionnaires.The analysis of the collected data based on the PET-USES (Privacy-Enhancing Technology Users Self-Estimation Scale) method [26], which considers usability aspects. The main research proceeding consists in solving identified problem areas. In this paper we focus on the crucial navigation problem (red in Fig. 6) for space reasons. Further information could be found in [27].

Since SocialTV mainly targets the integration of software/hardware to support social activities for either synchronous real-time interaction schemes or those of asynchronous nature, it presents a completely different scenario where new multi-user requirements related to media consumption have to be fulfilled while considering usability and privacy aspects.

Requirements Summary. Based on previously described work, a list of usability and security related requirements were identified. The main problem area is related to the navigation in a SocialTV environment in general. In summary, the designed solution should reflect realistic SocialTV situations (R1), allow for flexible parallel interaction of the involved people (R2), and be flexible in terms of costs emerging from adaptations to new situations and tests (R3), as well as allow for secure interaction should be supported thereby (R4).

To fulfill our requirements R1-R4 we extended the WallShare [28] platform. WallShare provides users with a collaborative multi-pointer shared desktop that is projected on a wall, or displayed on a big screen (R1, R2). Pointers are controlled through mobile devices, such as PDAs, smart-phones, tablet PCs, etc.

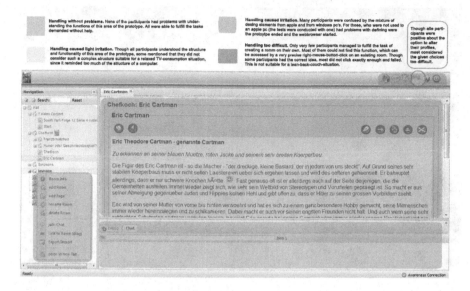

Fig. 6. First evaluation results

using dragging gestures over the mobile device screen (R2). Here, the mobile devices act as remote controls that can be adapted easily to new situations (R3). The system allows users to upload, or download resources to/from the shared desktop using a pointer that is controlled from users mobile devices by the means of gestures over the screen by considering access rights according to the Idemix anonymous credential system (R1-R4). Idemix [29] is a proof-based credential system, which could support such access to SocialTV adult content by only providing a proof, so that the person who is asking for such access has only to proof that he/she is over 18 years. In contrast to other credential systems, Idemix does not send attributes, which could lead to linkability (R4).

From the interactive perspective, according to the taxonomy described in [30], WallShare is a distributed user interface (DUI) represented by a multi-display ecosystem composed by Inch and Perch scale size displays that define a few-few social interaction relationship among users. From the collaborative perspective the system provides users with face-to-face collaboration, in the same space, at the same time (R3).

Fig. 7 shows a prototype realized by extending the WallShare platform where people navigate across the SocialTV environment by using their own mobile devices as remote controls are depicted. The system allows other groups to see the same content with this group together, which was required in our case to better study privacy aspects (e.g. listening people impressions while carrying out the tests). The description of the application is detailed in next sections.

However, one of the big needs consisted in earlier consideration of security and privacy in UI modeling especially due to frequent changes when prototyping

Fig. 7. SocialTV Scenario

the navigation. For this, we continue in the following with addressing how we proceed by adding security and privacy into UIs and we describe the way to keep tracing between layers throughout the mapping model.

5.2   Tasks and Concepts

The Task & Concepts layer of the framework is defined by two models: the domain model and the task model. While the domain model describes the concepts that will be manipulated by the SocialTV UI, the task models describe the interactions that will be carried out by the users of the system.

The notation we used to describe the domain model is the UML Class diagram. The domain model depicted in Fig. 8 describes most relevant parts of the SocialTV system regarding security and privacy issues. The SocialTV System is composed by a set of *Resources* and *Users*. While *Users* are defined by 4 attributes: *id, name, birthdate, subscription*, and *Resources* are described in terms of *name, ageRating* and *Subscription*. Note that although the attributes defined by *Resource* and *User* are related (*subscription, birthdate* and *ageRating*) there is no explicit relationship between users and resources. This relationship is defined in a separate model leveraging the level of reuse in the system.

The notation we used to describe the task model is CTT. The task model depicted in Fig. 9 describes the most relevant parts of the SocialTV system regarding security and privacy issues. Thus, a *SocialTVProcess* allows users to *NavigateOnResources* and *ManipulateVideo*. While the navigation task allows users to select a video by browsing on categories, the manipulation task allows users to control de video reproduction. As result of the analysis of this layer

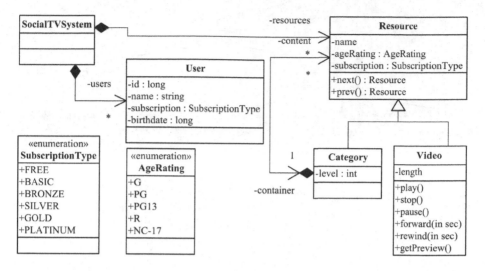

Fig. 8. SocialTV domain model

of models we can conclude that the domain model defines the data model of the system and the task model defines the behavior of the interaction system improving the decoupling of concerns at this level of abstraction.

5.3 Privacy and Security

Once the domain and task models are defined, next step is the definition of the privacy and security model. In order to define this model, we have to identify resources and the tasks that are affected by the security and privacy issues in the system to be developed. Thus, the resources that are affected are: *Category* and *Video*; and the tasks that are affected are: *ShowResourceItem*, *NextResource*, *PrevResource*, *Play*, *Forward* and *Rewind*.

Next, we have to define the *Principals* that will define the set of the users that will be able to interact with the system. Usually, you define at least two types of *Principals*, Users and Administrators. As we have mentioned in Section 4.1, Users and Administrators may employ different authentication policies. However, we define the same authentication methods for both Principals.

Last entities to be defined are the *SecurityAffordances*; in this case, we define the typical three levels of affordances INVISIBLE, DISABLED and ENABLED *SecurityAffordances*. We will mainly employ:

- INVISIBLE for tasks or resources that are reserved for adults. Thus, children are not able to seen inappropriate context.
- DISABLE for tasks or resources that are reserved for users with certain level of subscription. Thus, users are aware that they can get something more if they subscribe a higher level.
- ENABLE for tasks or resources than can be accessed without restriction.

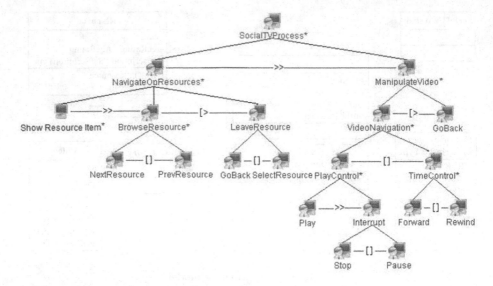

Fig. 9. SocialTV task model

After the definition of the entities that are part of the security and privacy model, we have to define the relationships among them. The first relationship to be defined is the *AccessRight*. The *AccessRight* defines those Attributes from the *Resource* and the *Principal* that are employed by the security and privacy process to calculate whether the *Resource* to be manipulated by a specified *Activity* is accessible or not. Besides, it defines for them the *SecurityAffordace* that the *Principal* perceives from the *Resource*.

Thus, we define the following *AccessRight* relationships: (User, Video, *user.birthday, user.subscription, video.ageRating, video.subscription*, ENABLED) and (User, Category, *user.birthday, user.suscription, video.ageRating, video.subscription*, ENABLED). Note that usually these relationships define the lowest *SecurityAffordace* for the resource; because they are refined by the *Access-Configuration* relationship. A higher level of affordance is used when there are Resources that cannot be used by some Principals. For instance, suppose that we include accounting module to the system. Only Administrators are authorized to access accounting resources.

The other relationship to define is the *AccessConfiguration*. It defines the relationship between Resources and Activities refining the *AccessRight* relationship by contextualizing the Activity that will be performed on the Resource with the appropriate *SecurityAffordance*.

The following tuples define *AccessConfigurations* for the *ShowResourceItem* Activity:

- (Video Under Age, ShowResourceItem, Video, [toAgeRating(User.birthday) < Video.ageRating], INVISIBLE)

- (Video Above Age, ShowResourceItem, Video, [toAgeRating(User.birthday) >= Video.ageRating], ENABLED)
- (Video Uncovered Subscription, ShowResourceItem, Video, [User.subscription < Video.subscription], DISABLED)
- (Video Covered Subscription, ShowResourceItem, Video, [User.subscription >= Video.subscription], ENABLED)
- (Category Under Age, ShowResourceItem, Category, [toAgeRating(User.birthday) < Category.ageRating], INVISIBLE)
- (Category Above Age, ShowResourceItem, Category, [toAgeRating(User.birthday) >= Category.ageRating], ENABLED)
- (Category Uncovered Subscription, ShowResourceItem, Category, [User.subscription < Category.subscription], DISABLED)
- (Category Covered Subscription, ShowResourceItem, Category, [User.subscription >= Category.subscription], ENABLED)

Where the first element of the tuple represents the name of the *AccessConfiguration* (used to identify it when performing the mapping process); the second element represents the *Activity*; the third element represents the *Resource*; the forth one represents the conditions to be satisfied in order to make the *Access-Configuration* valid; and fifth defines is the *SecurityAffordance* level.

Fig. 10 shows the security model for our SocialTV scenario addressed in this work based on the meta-model presented in Fig. 3.

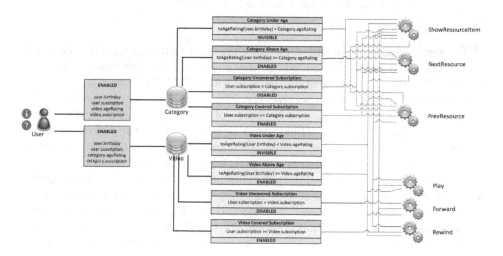

Fig. 10. SocialTV security model

5.4   Abstract User Interface

The abstract user interface model describes the UI in a model independent way. For instance, the same model can be used to describe a Graphical User Interface

GUI or a Vocal User Interface. The AUI is described in terms of containers and components that are characterized by facets (input, output control and navigation).

The SocialTV AUI model is depicted in Fig. 11 where we can see that the *SocialTVUI* is composed by two main containers: the *ResourceBrowser* and the *Video-Player*. The first one is in charge of providing SocialTV users the ability to browse information.

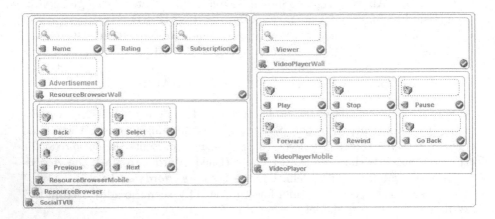

Fig. 11. SocialTV abstract user interface model

The second one is responsible for controlling the Video player. As the SocialTV system is defined as DUI, we added two containers on both, the *ResourceBrowser* and the *VideoPlayer*, representing the Wall and the *Mobile* device.

From the video player perspective, the *VideoPlayerWall* is composed by the Viewer output component that is in charge of playing the video on the shared surface; and the *VideoPlayerMobile* that is composed by the following set of control components: *Play, Stop, Pause, Forward, Rewind* and *GoBack* to control the video playback. And from the resource browser perspective, the *Resource-BrowserWall* is composed by 3 output components: *Name, Rating, Subscription* and *Advertisement* (key element to be mapped to the security model) to show *re-source* information on the shared display; and the *ResourceBrowserMobile* that is composed by two navigation components (*Next* and *Previous*) to navigate among categories sharing the same parent category, and two control components (*Back* and *Select*) to select a category or go back to the parent category, if any.

Thus, input, output, navigation and control facets of components can be linked to the security and privacy models in order to represent the customization of the UI components behavior.

5.5   Concrete and Final User Interfaces

Due to space reasons, the Concrete UI model will be discussed jointly with the Final UI model because, from the security and privacy perspective, there are no conceptual differences between them. The Final UI for the SocialTV system is defined as a Distributed User Interface (DUI) [31–33] based on the Model View Controller architecture where most of the view components are displayed on the shared screen, and the controller components are located on the mobile devices (see Fig. 12). To show how security and privacy issues are related to the CUI, we will define two users with different characteristics and will explain how the presence of users in the system affects the UI. In this case, for the sake of simplicity we will only show the effects on the shared display.

Suppose that we define two users: John (25 years old, silver subscription) and Peter (7 years old, no subscription). When Peter joins a SocialTV session, the shared screen shows the upper screen of the Fig. 12.

Lets analyze why the system is showing this screen (s. Fig. 12). It shows the Winnie the Pooh category because it is able to perceive that Peter is 7 years old and he is alone in the session. Besides, the system shows that Season 1 is the only season he is able to navigate because he has no subscription.

Now suppose that John joins the session. Thus, Peter is not alone anymore, and the system shows the lower screen on Fig. 12.

Lets analyze why the system has updated the screen. First, PG-13 movies can be watched by Peter because an adult is with him. Besides, as John has a Silver subscription they are able to see more seasons. The UIs depicted on Fig. 12 shows both UI properties that are linked to two UI issues: affordance and awareness. The affordance is expressed by showing the categories that are available according to the age of the user/s; and by showing the upgrade option

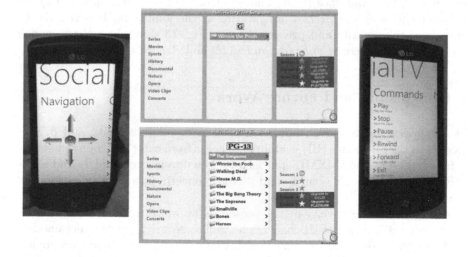

Fig. 12. SocialTV user interface: Shared UI in the center and Mobile UI on the sides

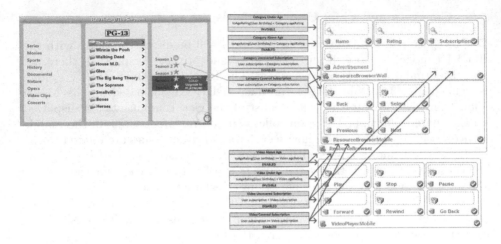

Fig. 13. Mapping model

on seasons that may be accessible if the user upgrades the subscription up. On the other hand, the awareness is expressed by disabling seasons.

5.6   Mapping

The mapping model is straightforward among elements at different models in the Task and Concepts layer of the Cameleon Framework. The actual mapping is done by linking the Resource at the Security model to the Resource defined at the domain model. Besides, we have to link the Task to Activities analogously.

However, is it interesting to note how the security and privacy models affect the Abstract User Interface (AUI) and the Concrete User Interface (CUI). As we have mentioned, the *AccessConfiguration* is the joint point between the UI models and the security and privacy models. Fig. 13 shows how these models are related by the *AccessConfigurationToAIO* and *AccessConfigurationToCIO*.

6   Conclusions and Future Work

In this paper a model-driven approach is used to relate security and privacy concerns to User Interface (UI) usability concerns from early development stages. We follow thereby the UsiXML approach for the development of UIs and adapt the PriS approach to add security and privacy concerns into the UI development in early stages.

We argue that the novelty of the presented approach consists in taking together the consideration of UI concerns as well as security/privacy metamodels. This is shown for the development of multi-platform and multi-modal secure and privacy enhancing UIs. We argued that this enforces the earlier consideration of such NFRs in the development. We suggested an extension of the UsiXML UIDL for addressing security and privacy issues in the UI metamodeling by presenting a premature security metamodel. The linkage of used models is carried out within a mapping model. For instance, for explicit authentication a password UI string field should not show the password and permit copy&paste. Another example is carrying out in other steps by using anonymous credential systems such as Idemix in the background without any interaction at the level of the UI.

This work goes beyond the contributions of related work since (1) approaches in the UI metamodeling and security research areas focus more on specific solutions in their own fields, and (2) the consideration of merging results from both fields is still not mature at metamodel level. With means of metamodeling we support the formulation and consideration of NFRs by allowing at the same time for the integration of multiple perspectives from multiple domains. With this, we just performed the first step and future work will target refining the security metamodel towards covering more privacy and security goals like unlinkability, unobservability, etc. Furthermore, the integrated support with the help of an Eclipse based plugin/editor should ease the work of the designer from the beginning.

Acknowledgments. This research has been partially supported by the Spanish CDTI research project CENIT-2008-1019 and the regional projects with reference PPII10-0300-4174 and PII2C09-0185-1030.

References

1. Limbourg, Q., Vanderdonckt, J., Michotte, B., Bouillon, L., Florins, M., Trevisan, D.: Usixml: A user interface description language for context-sensitive user interfaces. In: Proceddings of the ACM AVI 2004 Workshop Developing User Interfaces With XML: Advances On User Interface Description Languages, pp. 55–62. Press (2004)
2. Gross, T., Koch, M.: Computer-Supported Cooperative Workspace. Oldenburg (2007)
3. Shneiderman, B., Plaisant, C.: Designing the User Interface: Strategies for Effective Human-Computer Interaction, 4th edn. Pearson Addison Wesley (2005)
4. Chung, L., Nixon, B.A.: Dealing with non-functional requirements: three experimental studies of a process-oriented approach. In: ICSE 1995: Proceedings of the 17th International Conference on Software Engineering, pp. 25–37. ACM, New York (1995)
5. Woody, C., Alberts, C.: Considering operational security risk during system development. IEEE Security and Privacy 5(1), 30–35 (2007)
6. Kalloniatis, C., Kavakli, E., Gritzalis, S.: Addressing privacy requirements in system design: the pris method. Requir. Eng. 13(3), 241–255 (2008)
7. Cranor, L., Garfinkel, S.: Security and Usability. O'Reilly Media, Inc. (2005)
8. Shneiderman, B., Plaisant, C., Cohen, M., Jacobs, S.: Designing the user interface: strategies for effective human-computer interaction, 5th edn. Addison-Wesley Longman Publishing Co., Inc., Boston (2009)
9. Palen, L., Dourish, P.: Unpacking "privacy" for a networked world. In: CHI 2003: Proceedings of the SIGCHI Conference on Human Factors in Computing Systems, pp. 129–136. ACM Press, New York (2003)

10. Lukosch, S., Bourimi, M.: Towards an enhanced adaptability and usability of web-based collaborative systems. International Journal of Cooperative Information Systems, Special Issue on 'Design, Implementation of Groupware, 467–494 (2008)
11. Boyle, M., Neustaedter, C., Greenberg, S.: Privacy factors in video-based media spaces. In: Harrision, S. (ed.) n Media Space: 20+ Years of Mediated Life, pp. 99–124. Springer (2008)
12. Gutwin, C., Greenberg, S., Roseman, M.: Workspace awareness in real-time distributed groupware: Framework, widgets, and evaluation. In: Proceedings of HCI on People and Computers XI, pp. 281–298. Springer, London (1996)
13. Boyle, M., Greenberg, S.: The language of privacy: Learning from video media space analysis and design. ACM Trans. Comput.-Hum. Interact. 12(2), 328–370 (2005)
14. Miller, J., Mukerji, J.: Mda guide version 1.0.1. Technical report, Object Management Group (OMG) (June 12, 2003) (accessed August 24, 2011)
15. Wehrmeister, M., Freitas, E., Pereira, C., Wagner, F.: An aspect-oriented approach for dealing with non-functional requirements in a model-driven development of distributed embedded real-time systems. In: 10th IEEE International Symposium on Object and Component-Oriented Real-Time Distributed Computing, ISORC 2007, pp. 428–432 (May 2007)
16. Reichwald, J., Dörnemann, T., Barth, T., Grauer, M., Freisleben, B.: Supporting and Optimizing Interactive Decision Processes in Grid Environments with a Model-Driven Approach. In: Dolk, D., Granat, J. (eds.) Decision Support Modeling in Service Networks. LNBIP, vol. 42, pp. 14–35. Springer, Heidelberg (2012)
17. Jokela, T.: Assessment of User-Centred Design Processes - Lessons Learnt and Conclusions. In: Oivo, M., Komi-Sirviö, S. (eds.) PROFES 2002. LNCS, vol. 2559, pp. 232–246. Springer, Heidelberg (2002)
18. Schümmer, T.: A Pattern Approach for End-User Centered Groupware Development. Schriften zu Kooperations- und Mediensystemen - Band 3. JOSEF EUL VERLAG GmbH, Lohmar - Köln (August 2005)
19. Lieberman, H., Paterno, F., Wulf, V. (eds.): End User Development. Springer (2006)
20. Bourimi, M., Barth, T., Haake, J.M., Ueberschär, B., Kesdogan, D.: AFFINE for Enforcing Earlier Consideration of NFRs and Human Factors when Building Socio-Technical Systems Following Agile Methodologies. In: Bernhaupt, R., Forbrig, P., Gulliksen, J. (eds.) HCSE 2010. LNCS, vol. 6409, pp. 182–189. Springer, Heidelberg (2010)
21. Bourimi, M., Tesoriero, R., Villanueva, P., Karatas, F., Schwarte, P.: Privacy and security in multi-modal user interface modeling for social media. In: IEEE Third International Conference on Privacy, Security, Risk and Trust (PASSAT), and IEEE Third International Conference on Social Computing (Socialcom), pp. 1364 –1371 (October 2011)
22. Calvary, G., Coutaz, J., Thevenin, D., Limbourg, Q., Bouillon, L., Vanderdonckt, J.: A unifying reference framework for multi-target user interfaces. Interacting With Computers 15, 289–308 (2003)
23. Limbourg, Q., Vanderdonckt, J., Michotte, B., Bouillon, L., López-Jaquero, V.: USIXML: A Language Supporting Multi-path Development of User Interfaces. In: Bastide, R., Palanque, P., Roth, J. (eds.) EHCI-DSVIS 2004. LNCS, vol. 3425, pp. 200–220. Springer, Heidelberg (2005)

24. Vanderdonckt, J.M., Bodart, F.: Encapsulating knowledge for intelligent automatic interaction objects selection. In: Proceedings of the INTERACT 1993 and CHI 1993 Conference on Human Factors in Computing Systems, CHI 1993, pp. 424–429. ACM, New York (1993)
25. Paterno, F.: Model-Based Design and Evaluation of Interactive Applications, 1st edn. Springer, London (1999)
26. Bourimi, M., el Diehn I. Abou-Tair, D., Kesdogan, D., Barth, T., Höfke, K.: Evaluating potentials of Internet- and Web-based SocialTV in the light of privacy. In: 2010 IEEE Second International Conference on Social Computing (SocialCom), pp. 1135–1140 (August 2010)
27. Bourimi, M., el Diehn I. Abou-Tair, D., Kesdogan, D., Barth, T., Höfke, K.: Evaluating potentials of Internet and Web based SocialTV in the light of privacy. In: First International Workshop on Privacy Aspects of Social Web and Cloud Computing (PASWeb-2010) (2010) (in press)
28. Villanueva, P.G., Tesoriero, R., Gallud, J.A.: Multi-pointer and collaborative system for mobile devices. In: Proceedings of the 12th International Conference on Human Computer Interaction with Mobile Devices and Services, MobileHCI 2010, pp. 435–438. ACM, New York (2010)
29. Camenisch, J., Van Herreweghen, E.: Design and implementation of the idemix anonymous credential system. In: Proceedings of the 9th ACM Conference on Computer and Communications Security, CCS 2002, pp. 21–30. ACM, New York (2002)
30. Terrenghi, L., Quigley, A., Dix, A.: A taxonomy for and analysis of multi-person-display ecosystems. Personal Ubiquitous Comput. 13, 583–598 (2009)
31. Larsson, A., Berglund, E.: Programming ubiquitous software applications: requirments for distributed user interface. In: Proceedings of the Sixteenth International Conference on Software Engineering and Knowledge Engineering, SEKE 2004 (2004)
32. Luyten, K., den Bergh, J.V., Vandervelpen, C., Coninx, K.: Designing distributed user interfaces for ambient intelligent environments using models and simulations. Computers & Graphics 30(5), 702–713 (2006)
33. Luyten, K., Coninx, K.: Distributed user interface elements to support smart interaction spaces. In: Proceedings of the Seventh IEEE International Symposium on Multimedia, pp. 277–286. IEEE Computer Society, Washington, DC (2005)

Author Index